I0036548

Bionanotechnology for Advanced Applications

This book provides the fundamental aspects of bionanomaterials and bionanotechnology, and insight into the synthesis and modification of bionanomaterials in a detailed manner. It initiates with a general overview of biotechnology and nanotechnology followed by different strategies and methodologies for the synthesis of nanomaterials. Further, it discusses pertinent topics such as protein engineering, analysis, mechanisms of microbe-mediated nanosynthesis, followed by various challenges and innovation strategies, and the role of enzymes in bionanotechnology.

Features:

- Covers the synthesis of bionanomaterials, including the interaction between nanomaterial and biogenic materials.
- Encompasses the study of the connections between structure, molecular biology, and nanotechnology.
- Explains several techniques (XRD, SEM, TEM, etc.) used for the analysis of bionanomaterials.
- Includes prospects, challenges, and opportunities associated with bionanotechnology.
- Reviews the interaction between nanomaterials and the biological system and self-assembly in bionanotechnology.

This book is aimed at graduate students and researchers in materials sciences, biotechnology, and bionanotechnology.

Advances in Bionanotechnology

Series Editors: Ravindra Pratap Singh, *Department of Biotechnology, Indira Gandhi National Tribal University, Anuppur, Madhya Pradesh, India*, **Jay Singh**, *Department of Chemistry, Institute of Science, Banaras Hindu University, Varanasi, Uttar Pradesh, India* and **Charles Oluwaseun Adetunji**, *Department of Microbiology, Edo State University Uzairue, Iyamho, Edo State, Nigeria*

Bionanotechnology is a multi-disciplinary field that shows immense applicability in different domains, namely chemistry, physics, material sciences, biomedical, agriculture, environment, robotics, aeronautics, energy, electronics and so forth. This book series will explore the enormous utility of bionanotechnology for biomedical, agricultural, environmental, food technology, space industry, and many other fields. It aims to highlight all the spheres of bionanotechnological applications and its safety and regulations for using biogenic nanomaterials that are a key focus of the researchers globally.

Bionanotechnology Towards Sustainable Management of Environmental Pollution
Edited by Naveen Dwivedi and Shubha Dwivedi

Natural Products and Nano-formulations in Cancer Chemoprevention
Edited by Shiv Kumar Dubey

Bionanotechnology towards Green Energy: Innovative and Sustainable Approach
Edited by Shubha Dwivedi and Naveen Dwivedi

Biotic Stress Management of Crop Plants using Nanomaterials
Edited by Krishna Kant Mishra and Santosh Kumar

Bionanotechnology for Advanced Applications
Edited by Ajaya Kumar Singh and Bhawana Jain

For more information about this series, please visit: www.routledge.com/Advances-in-Bionanotechnology/book-series/CRCBIONAN

Bionanotechnology for Advanced Applications

Edited by
Ajaya Kumar Singh and Bhawana Jain

CRC Press
Taylor & Francis Group
Boca Raton London New York

CRC Press is an imprint of the
Taylor & Francis Group, an **informa** business

First edition published 2024
by CRC Press
2385 NW Executive Center Drive, Suite 320, Boca Raton FL 33431

and by CRC Press
4 Park Square, Milton Park, Abingdon, Oxon, OX14 4RN

CRC Press is an imprint of Taylor & Francis Group, LLC

© 2024 selection and editorial matter, Ajaya Kumar Singh and Bhawana Jain; individual chapters, the contributors

Reasonable efforts have been made to publish reliable data and information, but the author and publisher cannot assume responsibility for the validity of all materials or the consequences of their use. The authors and publishers have attempted to trace the copyright holders of all material reproduced in this publication and apologize to copyright holders if permission to publish in this form has not been obtained. If any copyright material has not been acknowledged please write and let us know so we may rectify in any future reprint.

Except as permitted under U.S. Copyright Law, no part of this book may be reprinted, reproduced, transmitted, or utilized in any form by any electronic, mechanical, or other means, now known or hereafter invented, including photocopying, microfilming, and recording, or in any information storage or retrieval system, without written permission from the publishers.

For permission to photocopy or use material electronically from this work, access www.copyright.com or contact the Copyright Clearance Center, Inc. (CCC), 222 Rosewood Drive, Danvers, MA 01923, 978-750-8400. For works that are not available on CCC please contact mpkbookspermissions@tandf.co.uk

Trademark notice: Product or corporate names may be trademarks or registered trademarks and are used only for identification and explanation without intent to infringe.

ISBN: 9781032416120 (hbk)
ISBN: 9781032423227 (pbk)
ISBN: 9781003362258 (ebk)

DOI: 10.1201/9781003362258

Typeset in Times
by Newgen Publishing UK

Contents

Preface

The present book, *Concepts of Bionanotechnology for Advanced Applications*, deals with a subject of high interest and importance on the fundamentals of bionanotechnology. It is the first book of its kind, which covers all aspects bionanotechnology from start to finish. It has been written collaboratively by biologist and chemists. Biological systems are essential in nanotechnology, and many new applications are being developed by mimicking natural systems. Approaching these topics from an engineering perspective, the book offers an insight into the details of nanoscale fabrication processes as well as cell biology. The basics of biology and chemistry, with a focus on how to engineer the behavior of molecules at the nanoscale, are also explored and analyzed. The aim of this text is to provide the reader with a broad knowledge of the biological methods for signal transduction and molecular recognition systems, and how they can be replicated in bio-sensing applications. The reader will learn the basic structures and interactions of biomacromolecules.

This book also explains the basics of bionanomaterial properties, synthesis, and chemistry, and demonstrates how to use bionanomaterials in diverse fields to overcome problems in agricultural, environmental, and biomedical areas. These materials play a key role in everyday life and will have an important effect on the future of technology. Therefore, exciting developments, new advancements and foundations, and directions for further exploring bionanotechnology are necessary to build on the field's current status.

This book is an outstanding collection of current research on basic and advanced bionanomaterials, providing an overview of selected advances that have taken place in the field of nanotechnology. Moreover, this book introduces the basic concept of the nanotechnology, bionanotechnology, the interaction of biological molecules with nanomaterials, and its advanced applications to help students and professionals better understand these materials. This book will serve as a reference book for professionals, students, scientists, researchers, and academicians in this subject area.

Finally, we hope that readers will find it beneficial to their future research and teaching endeavors.

AIMS AND SCOPE

Concepts of Bionanotechnology for Advanced Applications will attract a wide range of readers from all fields as it will provide the fundamental aspects of bionanomaterials and bionanonotechnology, with an excellent insight into the synthesis and modification of bionanomaterials. Additionally, this book will also provide a single platform for the scientific community working on different aspects of bionanotechnology. Our ambition is for this book to offer highlights of new contributions for experienced researchers and provide an introduction to this fascinating research field for novices.

ORGANIZATION OF THE BOOK

The 11 chapters in the book provide an insight into the synthesis and properties of bionanomaterials and bionanotechnology in detail. We hope that readers will enjoy reading this book as much as we have enjoyed putting it together.

Chapter 1 defines nanotechnology and bionanotechnology for various materials. This chapter is an overview of the emerging nanotechnology and related dimensions in terms of its applications, challenges in hand, and a future perspective.

Chapter 2 deals with different biological systems and bionanotechnology for various fields. Bionanotechnology is an extension of nanotechnology that utilizes biology for technological applications at the nanoscale.

Chapter 3 includes a description of the fabrication of nanomaterials via different techniques including different green sources as well as chemical methods, along with some physical routes such as ball milling.

Chapter 4 describes nano-fertilizer development by enclosing plant nutrients in nanomaterials and distributing them in the form of nano-sized emulsions. Nano-biosensors are nanomaterials with characteristics that improve sensing mechanisms. Nanotube-based sensors, nanoparticle-based biosensors, nanowire-based sensors, and quantum dot-based sensors are examples of biosensors. Several nanoparticles (Ag, Fe, Cu, Al, Si, Zn, ZnO, TiO_2, CeO_2, Al_2O_3, and carbon nanotubes) have been identified as having negative impacts on plant growth, bringing the use of nanoparticles in agriculture into question. Thus, the role of nanotechnology along with biology in the agriculture field is demonstrated in Chapter 4.

Chapter 5 describes the application of microbial enzyme nanoparticles for biological regulation and metabolism that is swiftly increasing due to growing strains in the field of environmental monitoring, biochemical engineering, and biomedicine.

Chapter 6 describes the role of nano-technology in the fields of disease diagnostics, therapeutics, antimicrobial, anti-biofilm agents, drug/vaccine delivery systems, food safety, air and water cleansing, waste management, and disinfection of veterinary care instruments/implants. Hence, the role of nanobiotechnology in providing a sterile and safe environment for the animal and the human environment is discussed in detail.

Chapter 7 describes bionanotechnology, a branch of science, which is explored for monitoring toxic analytes from food and cosmetic products. The presence of multi-functional sites in biomolecule-functionalized nanomaterials provides a surface for capturing toxicants. It can be helpful for the design and development of new portable materials that could be used in remote areas for public safety.

Chapter 8 describes forensic nanotechnology minute chip materials that are used, rather than bulky instruments, to enhance the methods investigation to be more accurate, precise, timely, and appropriate. This chapter spotlights a number of applications of nanotechnology in forensic science.

Chapter 9 describes the advantages of several recent microbial approaches in the field of nanoparticle synthesis. In addition, future perspectives of nanoparticles are also incorporated.

Chapter 10 describes the application of nanotechnology in the development of herbal formulations and in the treatment of various disease conditions such as cancer.

Chapter 11 describes nanobiotechnology as a relatively new field that will take some time to gain the public trust; the potential it has in agriculture and allied fields is immense for improving crops, crop management, agriculture waste treatment and management, as well as crop disease management and crop protection, in turn increasing productivity.

Contributors

Akshay Bharti
School of Life and Basic Sciences, Jaipur National University, Jaipur, India

Anand Pithadia
School of Pharmacy, Parul University, Gujarat, India

Anil Jogdand
National Centre For Cell Science, Pune, India

Smita Badur Karmankar
Department of Chemistry, IPS Academy – Institute of Engineering and Science, Rajendra Nagar Indore, India

Bhavisha Patel
School of Pharmacy, Parul University, Gujarat, India

Brij Mohan
College of Ocean Food and Biological Engineering, Jimei University, Xiamen, China

Chandni Krishnan
School of chemical and Biotechnology Sastra Deemed University, Tirumalaisamudram, Thanjavur, India

Chetna Sharma
Department of Applied Chemistry, Aligarh Muslim University, Aligarh, India

Ranjana Ahirwar Choudhary
Department of Chemistry, IPS Academy – Institute of Engineering and Science, Rajendra Nagar Indore, India

Darshini Trivedi
Department of Botany, Faculty of Science, The Maharaja Sayajirao University of Baroda, Sayajigunj, Vadodara, Gujarat

Divakar Sharma
Research Scientist, Department of Microbiology, Lady Hardinge Medical College, Shaheed Bhagat Singh Marg, New Delhi, Delhi, India

Shradhha Dwivedi
College of Veterinary & Animal Sciences, Govind Ballabh Pant University of Agriculture & Technology, Pantnagar, Uttarakhand, India

Ekta Poonia
Department of Chemistry, Deenbandhu Chhotu Ram University of Science & Technology, Murthal, Sonepat, Haryana, India

Esha Rami
Department of Life Science, Parul Institute of Applied Sciences, Parul University, Vadodara, Gujarat, India

Arti Hadap
Basic Science and Humanities Department, MPSTME, NMIMS Mumbai, Maharashtra, India

Harsha Devnani
Department of Sciences, School of Sciences, Manav Rachna University, Faridabad, India; University Instrumentation Center, Manav Rachna University, Faridabad, India

Pradip Hirapure
Rajiv Gandhi Biotechnology Center, Laxminarayan Institute of Technology Campus, Rashtrasant Tukadoji Maharaj Nagpur University, Nagpur, India

Bhawana Jain
Siddhachalam Laboratory, Raipur, India

Jaya Tuteja
Department of Sciences, School of Sciences, Manav Rachna University, Faridabad, India

Jude Juventus Aweya
College of Ocean Food and Biological Engineering, Jimei University, Xiamen, China

Juhi Sharma
School of Life and Basic Sciences, Jaipur National University, Jaipur, India

Karan Sharma
School of Allied Health Sciences, Jaipur National University, Jaipur, India

Swati Mehra
Department of Chemistry, IPS Academy – Institute of Engineering and Science, Rajendra Nagar Indore, India

Anupriya Misra
College of Veterinary & Animal Sciences, Govind Ballabh Pant University of Agriculture & Technology, Pantnagar, Uttarakhand, India

Murthy Chavali
Dr. Vishwanath Karad MIT World Peace University (MIT-WPU), Kothrud, Pune, Maharashtra, India

Neeta Gupta
Govt. E. Raghavendra Rao P.G. Science College, Bilaspur, India

Pracheta Salunkhe
Department of Life Sciences, University of Mumbai, Maharashtra, India

Priya Choudhary
School of Life and Basic Sciences, Jaipur National University, Jaipur, India

Quansheng Chen
College of Ocean Food and Biological Engineering, Jimei University, Xiamen, China

Richa Das
Department of Life Science, Parul Institute of Applied Sciences, Parul University, Vadodara, Gujarat, India

Roopa Rani
Department of Sciences, School of Sciences, Manav Rachna University, Faridabad, India

Sarabjot Kaur Makkad
Department of Chemistry, Govt NPG College of Science, Raipur, C.G., India

Arti Shanaware
Rajiv Gandhi Biotechnology Center, Laxminarayan Institute of Technology Campus, Rashtrasant Tukadoji Maharaj Nagpur University, Nagpur, India

Alka Sharma
Department of Chemistry, IPS Academy – Institute of Engineering and Science, Rajendra Nagar Indore, India

Shreni Agrawal
Department of Life Science, Parul Institute of Applied Sciences, Parul University, Vadodara, Gujarat, India

Surabhi Sharma
School of Allied Health Sciences, Jaipur National University, Jaipur, India

Anuj Tiwari
Department of Veterinary Microbiology, College of Veterinary & Animal Sciences, Govind Ballabh Pant University of Agriculture & Technology, Pantnagar, Uttarakhand, India

Princi Tiwari
College of Veterinary & Animal Sciences, Govind Ballabh Pant University of Agriculture & Technology, Pantnagar, Uttarakhand, India

Shreya Tiwari
College of Veterinary & Animal Sciences, Govind Ballabh Pant University of Agriculture & Technology, Pantnagar, Uttarakhand, India

Diksha Upreti
College of Veterinary & Animal Sciences, Govind Ballabh Pant University of Agriculture & Technology, Pantnagar, Uttarakhand, India

Vijay J. Upadhye
Center of Research for Development (CR4D) and Department of Microbiology, Parul
    Institute of Applied Sciences (PIAS), Parul University, Vadodara, Gujarat, India

Vijay Upadhye
Research and Development Cell (RDC) and Department of Microbiology, Parul
    Institute of Applied Sciences (PIAS), Parul University, Vadodara Gujarat, India

About the Editors

Ajaya Kumar Singh serves as a Professor in the Department of Chemistry at Govt. V.Y.T. PG Autonomous College, Durg (C.G.), India. He earned his Ph.D. in kinetics and oxidation from the University of Allahabad, Uttar Pradesh, India. Dr. Singh recently assumed the role of a visiting professor at the School of Chemistry & Physics, University of KwaZulu-Natal, Durban, South Africa.

His research spans diverse areas, encompassing the development of strategies involving synthesis, characterization, properties modification, design, fabrication, and characterization of nanostructures, quantum dots, perovskite materials, and photocatalysts. These efforts are directed toward applications in wastewater treatment and advanced oxidation processes. Dr. Singh boasts an impressive publication record, with over 170 contributions, including book chapters and articles in peer-reviewed journals. His works have garnered over 7000 citations, and he holds an H-index of 30.

Under Dr. Singh's guidance, one post-doctoral researcher and 17 doctoral students have received their respective degrees. His significant contributions to the field have been recognized through various awards. As an active member of numerous professional societies, he also serves as a reviewer for several peer-reviewed journals.

Dr. Singh has completed various state- (CCOST) and national-level (DST/UGC) research projects. Currently, he is involved in an international project as part of the India–Bulgaria Bilateral Exchange Program. His research has played a pivotal role in advancing the fundamental understanding of nanomaterials for wastewater treatment.

Bhawana Jain received her doctorate degree in year 2011 from Govt. V.Y.T.PG. Autonomous College, Durg (C.G.), India. She has been awarded prestigious fellowships from UGC (Postdoctoral Fellowship for Women) and worked as a post-doctoral fellow at Govt. V.Y.T.PG. Autonomous College, Durg (C.G) India. She has 15 years of research experience. She is actively engaged in the development of nanomaterial-based (CeO_2, graphene oxide, ZnO, CuO, NiO, etc.) nanobiocomposites for treatment of wastewater containing different pollutants, dyes, drug, organic matter, etc. Dr. Bhawana Jain has published 30 international research papers and 10 book chapters, with a total of citations at <300, and her h-index being 10. Her research work findings also have been presented in various national and international (United States, Netherlands) platforms, such as conferences/seminars/symposia, etc. She received the Certificate of Merit from the American Chemical Society, United States, in 2016. She is a lifetime member of the Indian Science Congress and an annual member of the American Chemical Society. She has served as a reviewer for the publisher Taylor & Francis. Currently, she is actively engaged in the research and development field at Siddhachalam Laboratory as a founder.

1 Nanotechnology and Bionanotechnology
An Overview

Jaya Tuteja, Roopa Rani, Chetna Sharma, and Harsha Devnani

1.1 INTRODUCTION

The term "nano" found its inspiration from Richard Feynman's lecture, "There's plenty of room at the bottom," at the annual meeting of the American Physical Society held at the California Institute of Technology. At the time, Feynman's words made sense as per theory, but its practicality seemed a distant reality [1]. He elaborated the possibilities of working at the atomic or molecular level beyond the limitations of theoretical physics. He was a visionary in the sense that he predicted the advent of time wherein precise manipulation of matter could be realized. His lecture thus marks one of the first landmarks in the building of nanotechnology. The prefix "nano" has its origin in the Green word "nanos" meaning "dwarf" [2]. In the present terminology, the word "nano" indicates 10^{-9} of a measuring unit, and nanotechnology is the science, fabrication, and research at the nano scale of 1–100 nm. Nanomaterials are materials which possess a minimum of one dimension in the nanoscale range. Although more commonly "nano" refers to a size less than 100 nm, this definition is not as stringent when we are dealing with biomedical applications where the particles can range in size up to 1000 nm [3]. Nanoscience is not limited to working at the nanoscale, rather in a broader perspective it is about understanding and making use of changes in properties that are observed at the nanoscale. There is a transformation of mechanical, optical, thermal, electrical, physical, and chemical properties as we move from the macro to the nano world. There are significant changes at the nanoscale which are attributed to the dominance of quantum effects at this size scale affecting material properties.

Nanomaterials and nanotechnology revolve around designing based on their dimensions and forms. One can think of these as simply particles, tubes, wires, films, flakes, or shells with one or more dimensions measuring in nanometers. Nano-textured surfaces, for instance, have their thickness (one dimension) on the nanoscale, whereas nanotubes have two dimensions (the diameter of the tube) in the range of 0.1–100 nm, while the length can go further [4]. Nanofilms or nanoplates can have multiple aspects that are much larger than their nanoscale depth. Ultrafine particles (UFPs) is also popularly used synonymously with nanoparticles although the former

DOI: 10.1201/9781003362258-1

can be in the range of micrometers. Ultrastructure is a more common terminology when referring to biological nanostructures.

For many years, biologists have worked at the molecular level, with objects varying in size from nanometers (DNA and proteins) to 1 µm (cells). The double spiral of DNA is roughly 2 nm wide, a typical protein like an erythrocyte has a dimension of about 5 nm, and a mitochondrion is a few hundred nanometers long. Because of this, the study of any subatomic organism might be referred to as "nanobiology." The live cell is also regarded as the ideal nanoscale fabrication contemporary moment because of the millions of nano machines it contains [5]. In order to study and alter biological systems, nanotechnology provides the technological platforms and tools, and biologists supply nanotechnology with model organisms and bio-assembled parts. The word's technological component is what distinguishes "nanobiology" from "nanobiotechnology." Technology in the context of nanobiotechnology refers to everything "man-made." Considering biological systems include analogs to nearly every piece of cellular mechanism we can imagine, it looks like the first truly innovative uses of nanobiotechnology will likely be in the information and medical fields. In a sense, the creation of recombinant proteins could be the first application of a nanobiotechnology system [6]. With the use of genetic modification, specific polypeptides that may be used as parts of larger molecular structures can be produced both in vivo and in vitro by the ribosomal complex. Large materials are processed into fine structures with distinct surface features using optical and electron beam lithography as one of the technologies. The fact that the primary method in nanotechnology is top-down, while the dominating approach in nanobiotechnology is bottom-up, is one of the key distinctions between the two fields. The groundbreaking research of two significant groups on biomolecule microfilaments is an illustration of the bottom-up method [7,8]. In these investigations, naturally existing mobile enzymes were modified to be friendly with synthetic interfaces in order to devise novel strategies for attaching proteins to artificial nanomaterials. Micro- and nanotechnology can be moved chemically using biomolecular motors. Conveyor peptides like kinesin or F1-ATPase can be utilized in nanodevices to control the motion of an analyte, act as probes for surface imaging, and facilitate the controlled assembly of nanomaterials. The geometric qualities that enable DNA to function effectively as mitochondrial DNA can also be utilized to create bottom-up target materials with regular three-dimensional (3D) shapes. The utilization of crystalline microbial cells interface level (S-layers) proteins as tools for nanoscience and nanotechnology and nanopatterning is an additional noteworthy example [9]. The outermost layer of many bacteria's cell envelope, the S-layer, is made up of identical amino or glycosyl subunits that self-assemble into lattices. S-layers are intriguing model systems for investigations on the geometric, functional, and dynamic aspects of supramolecular structure formation because of their high degree of functional regularity. Additionally, nanoparticles of Au and CdSe were directly placed onto the protein lattice. Given that the interparticle distance and geometry affect the macroscopic electronic or magnetic properties of nanoscale panels, it should be possible to use different natural or artificial S-layer lattices as a "tunable" system to produce nanoparticle assemblies with desired properties for nanomaterials. These arrays of metallic or nanocrystals will serve as the building blocks for materials with specific electrical or magnetic characteristics. S-layers have been proposed for usage in a variety of applications, including as immobility matrices for biomedical applications or as templates for the nanoscale tailoring of inorganic compounds. Moreover, S-layer innovations in particular offer novel ways for nanoscale materials, surface nanopatterning, bioengineering, biomimetic, and the generation of ordered arrays of metal clusters or nanoscale needed for nanodevices [10]. There are a few techniques for creating nanodevices that could potentially be incorporated into living things without running the risk of being rejected as antigens. Another technique is to create nano- to nanostructures using living "tools." Each and every biological system is built on a foundation of nanoscale molecular components and machinery that work together to create living things. The fact that the primary method in nanoscale is top-down, while the dominating approach in biomaterials is bottom-up, is one of the key distinctions between the two fields.

1.2 OVERVIEW OF A WIDE RANGE OF APPLICATIONS

The field of nanotechnology is extensive and aims to integrate the interdisciplinary sciences at the nano level for a range of applications. There could be a chapter on each of the below-listed applications but here the authors attempt to give an overview of some of the major application fields and how nanotechnology is making a difference there. While nanotechnology has matured over the years, nanobiotechnology remains in the developmental stages but is multidirectional and is gaining impetus with time as the near future demands nanostructured devices based on biological molecules for an array of technological advancement. The bio-inspirational models in adjunction with the tools of nanotechnology pave the way for designing of unique nanofabricated devices and machinery in the near future (Figure 1.1) [11].

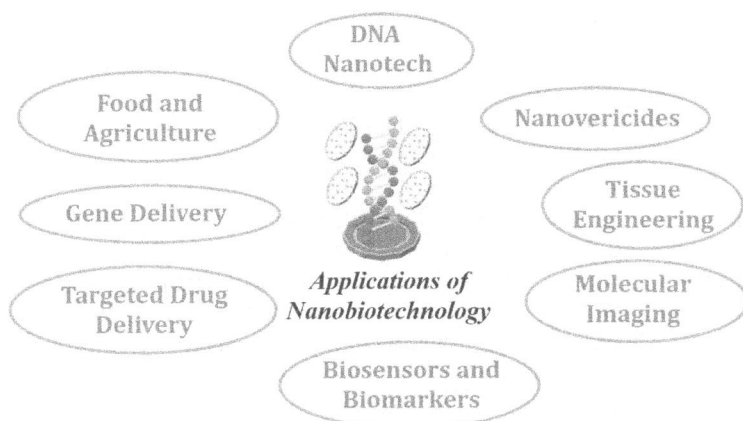

FIGURE 1.1   Schematic of nanobiotechnology applications.

1.2.1   MEDICAL AND HEALTHCARE

It is widely acknowledged that advancements in technology and medicine have significantly enhanced improved the life span of human. Nanotechnology has already gained a significant place in the treatment of life-threatening diseases in the form of nanomedicines; if we look further into this we will see that not just treatment, but also the right diagnosis of diseases is dependent on nanotechnology and nanobiotechnology. Nanotechnology also is specifically being used for targeted delivery to a larger extent. In the medical world, the first ever nanosystem was employed for nanomedicines, with the target of increasing drug efficacy, targeted site-specific delivery, and minimal side effects [12]. Nanoparticles (NPs) have found extensive application as therapeutic agents and have been widely studied for numerous preclinical and clinical investigations, which can also be seen by the variety of scientific reports and patents filed in this particular area. NPs possess a high surface area which is suitable for high drug loading, they are very small in size which ensures their ability to cross a variety of barriers in the body, and the third and most important quality is the tunable surface charge which means we can easily alter the surface charge to make it possible to pass through cellular membranes [13–16]. Owing to these advantages, NPs are considered as promising agents for drug delivery.

A variety of polymeric, hydrophilic, and hydrophobic vehicles can be designed or synthesized by the assembly of a copolymer chain of numerous shapes and sizes [17]. The resulting polymeric NPs are suitable to incorporate both hydrophobic and hydrophilic drug moieties and are also capable of incorporating macromolecules of proteins, nucleic acids, and antibodies [18]. Recent researches have highlighted the importance of pH-sensitive NPs for the treatment of cancer or tumor cells. These pH-sensitive NPs are capable of differentiating tumor tissue cells (pH 5.7–7.0) from normal tissue cells (pH 7.4) and thus show high site-specific targeted drug delivery and minimal side effects to normal tissue growth [19].

1.2.2   ELECTRONICS AND IT APPLICATIONS

As in the medical and healthcare industries, nanotechnology is known for site-specific and high dosage of drug delivery, in electronics and IT applications nanotechnology is known for faster and portable systems. Here we discuss another important term known as "nano magnetics," which is a combination of nanotechnology with electronics/magnetics. If we just look around ourselves, we will notice that the size of transistors has been reduced significantly in recent years [20]. The smaller size and faster performance are achieved by controlling the properties of materials at the nanoscale. For instance, carbon nanotubes have been investigated as cooling microprocessor chips in the United States. Microprocessor units or CPUs, which are the heart of any computer, work on a chip which is generally densely designed to make it perform faster and be portable. Due to dense packing, the chip shows symptoms of overheating, thus to protect the chip from overheating, carbon nanotubes are used as thermal transistors. It has been concluded that carbon nanotubes exhibit high electrical conductivity and high thermal resistance with the disadvantage of high cost. Research has been continuously conducted in this area to solve the concern about high cost.

Another milestone achieved with the help of nanotechnology is MRAM (magnetic random access memory) which enables a system to start and restart in a fraction of a second. The MRAM works on the basis of magnetic tunnel junctions at the nano scale which make it possible to save data swiftly and efficiently during the reboot process [21].

The latest development of LG OLED roll on TV is another example of nanotechnology implications in electronic devices. These flexible, bendable transistors for TV application, medical appliances, and IoT applications are made by semiconductors of nanomembranes. For instance, graphene and cellulose nanomaterials have been used to make flexible rollable electronic devices. Another milestone achieved in this area is quantum dots, which are known as nanocrystals and have the capacity to transport electrons [22]. Quantum dots emit colorful lights when UV light strikes their surface. These quantum dots are artificial semiconductors that are being extensively used to provide high-definition display and bright colors on televisions.

1.2.3 NANOTECHNOLOGY IN ENERGY APPLICATIONS

Energy is the need of the era owing to the depletion of fossil fuel resources and the exponential rise in population growth and demand. There is a strong urge to develop new energy sources based on renewable resources. Among renewable energy sources, biomass is the one with C energy stored within, which can be further utilized to develop alternate fuel precursors and ultimately reduce the load on fossil fuel resources. For this transformation of biomass to fuel precursors the use of nanotechnology and nanobiotechnology as catalysis has been seen to play a vital role. Carbon metal nanocomposites [23], polymer nanocomposites [24], metal and metal oxide nanoparticles [25], and enzyme mimic nanocatalysts [26] are a few examples which have shown tremendous improvements in catalytic transformation of biomass.

The efficacy of solar panels can also be increased with the help nanotechnology by increasing the storage capacity of solar energy followed by quick conversion into electric energy. The first-generation solar cell uses the multicrystal system of size 300–400 microns [27]. The advancement of this system leads to the formation of thin films of polymer or metal semiconductors, also classified as second generation [28]. The third-generation solar systems include a print-like methodology or dye-sensitized solar cell which means instead of panels flexible roller sheets are used. Researches are still ongoing to make paintable solar cells which would be very helpful [29]. Nanotechnology has made significant contributions to the field of electric batteries, enabling advancements in rapid charging, extended battery life, and reduced weight [30].

1.2.4 NANOBIOTECHNOLOGY IN THE AGRICULTURE FIELD

Sustainable Development Goal 3 (SDG 3) has the aim to: "End hunger, achieve food security and adequate nutrition for all, and promote sustainable agriculture." This means that high-quality nutritious food for all should be the first priority. The challenging task in achieving this goal is the low productivity and huge losses in the agriculture sector owing to natural resource degradation by climate or environmental variations and many other factors. Farmers all over the world are looking for new innovations and technologies from scientists to produce high-quality crops with good nutritional value. Current research is focused on the developments of nano-based stimulants for precision farming. Nanotechnology has helped to improve soil quality, water retention, nutrient supply, safety in distribution of food, and food preservation [31] (Figure 1.2).

The biggest challenge in the agriculture field is the management of resources, and with the advancement of time it has been seen that nanotechnology has helped a great deal in the agriculture field in regard to precision farming, gene technology for DNA transfer in plants for specific qualities, reduced amount of pesticides by using nanomaterials, developing new plant varieties, and much more. Nanotechnology has played a vital role in improving soil quality and water retention owing to its unique material properties. Nanomaterials possess the property of being porous, with a high surface area and high absorbing capacity, and these qualities make them useful in the agriculture field for supplying nutrients to soil, controlled release of water as per requirements, and possessing a high amount of nutrients by encapsulation [32].

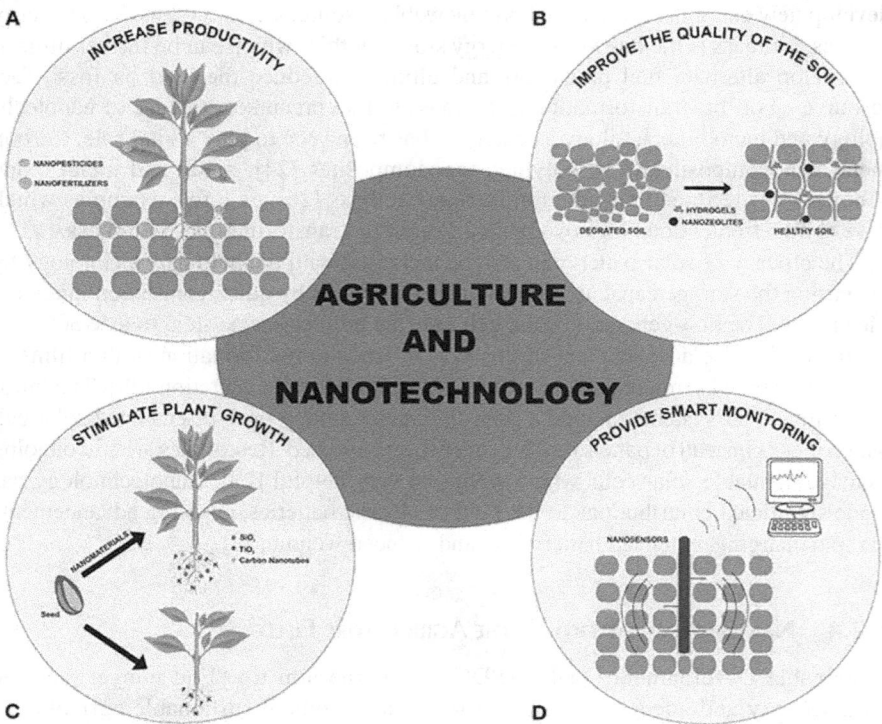

FIGURE 1.2   Nanotechnology's role in agriculture.

Source: [33].

1.3 TOOLS AND TECHNIQUES FOR NANOTECHNOLOGY

Scientists believe that the introduction of the environmental scanning electron (ESE) and scanning tunnel microscope (STM) in the 1980s opened up the field of nanotechnology. By rolling a tip over through the surface and analyzing the electron channeling or related forces, these microscopes can image a surface effectively. Moreover, the STM can be used to impact atoms on a surface, such as to build a quantum well. A broad range of scanning probe microscopes are currently available that can image a variety of material characteristics, such as magnetic probe, electromagnetic properties, etc.

A comprehensive and precise analysis of the material is crucial in order to examine the unique properties of nanoparticles. The conventional technologies reached an impasse in analyzing the unique properties of nanoparticles as they could not probe at the nanoscale. To deal with the problem of characterization of nanostructures, various tools and methods were created. Scanning electron microscopy (SEM), field emission scanning electron microscopy (FESEM), tunneling electron microscopy (TEM), atomic force microscopy (AFM), and other techniques and technologies were developed to analyze and image the morphology of nanomaterials. Using a probe to touch the sample's surface, the effective technique of AFM gathers data. It can be used to manipulate the surface of the sample to acquire high-resolution photographs on a variety of solid surfaces, and measure surface forces [34,35]. The particle tracking velocimetry (PTC) approach, which uses image-based velocimetry, is another cutting-edge technology for tracking individual particles in fluidic systems [36,37]. It involves photographing within a predetermined region where the particles of interest are lit utilizing fluorescent materials as tracers. By identifying the particles in those images, the motion patterns of the particles may be reconstructed, and their velocities can be computed in accordance. On the basis of this, profound understanding of some of the intricate and slow-moving flows in a zone can be gained. By using electrons as a "light source," which have a way shorter frequency than light, TEM could produce images with a great deal greater resolution than a light microscope [38]. A picture is created when an electron beam traverses a very thin sample and interacts with some of the sample's atoms before reaching a fluorescent screen. Different shade levels are used to present the image, illustrating how dense the substance is throughout the specimen. It is possible to study the image immediately from the screen or record it with a camera for later examination. Real-time monitoring of the phase transformation of nanostructures in response to active external stimuli and how such responses relate to attributes is possible with in situ TEM [39]. Numerous optical measures, particularly spectroscopic methods, are used to evaluate the optical properties of nanostructures. UV-visible spectroscopic, X-ray photoelectron spectroscopy (PL), Raman spectroscopy, and transmittance spectroscopy, are among the primary evaluation methods. The agglomerate of nanoparticles can be studied using UV-visible spectroscopic techniques. Getting the study material's absorption band edge is another useful thing to do. Using the appropriate equations, the particle size of the system may likewise be determined from the spectrum [40]. An analysis of a nanomaterial's emission spectrum gives clues about its structural flaws. With the advancement of sophisticated technologies such as SEM, PL, AFM, STM, and TEM, researchers have been capable of creating views at the nanotechnology level since the 1930s.

The ability to directly analyze the intracellular signaling complex using optical techniques, such as confocal fluorescence microscopy or correlation imaging, can be achieved by labeling target molecules with quantum dots (QDs) or synthetic chromophores, such as fluorescent proteins [41,42]. Nanotechnology can be used to create novel medicine formulations with fewer negative effects and drug distribution channels. There seem to be numerous cases of complex interactions between molecules and surfaces in nature. For instance, extensive interactions between cells and surface properties are necessary for the interactions between blood cells and the brain or between fungal pathogens and infection sites. By manipulating surface properties at nanoscale resolutions, nanofabrication reveals the complexity of these interactions, which can evolve in hybrid biological systems. The resulting material has potential uses in medical devices, implants, sensors, and random drug testing. A polymer coating created by Nano Systems, a subsidiary of the Irish drug company Elan, can alter the surface of medications with poor water solubility [43].

Nano manufacturing is the term for producing at the nanoscale. Nanoscale materials, structures, devices, and systems are manufactured at scales that are dependable, inexpensive, and efficient. Secondly, it involves the study, design, and integration of bottom-up, or consciousness, processes as well as top-down processes that are growing increasingly complex. Nanotechnology encompasses a wide range of topics, from modifications of standard device physics to entirely novel strategies based on molecular consciousness, from creating novel materials with nanoscale size to examining if we can directly control things at the atomic level. The implementation of this idea involves the use of numerous scientific disciplines, including surface science, chemistry, molecular genetics, photovoltaic cells, nanofabrication, etc.

1.4   ETHICS OF NANOTECHNOLOGY

There are several consequences of the use of nanotechnology with respect to the ethical issues like the contravention of privacy, infringement of a person's sovereignty, pollution of environment, exploitation of finances, safety and honesty, etc. The major implications of the use of nanotechnology are concern with the uncertainty and danger resulting through the manufacture of materials at the nano scale. Insecurity hitches in the commodification of nanotechnology can cause threat to human health and also raise environmental concerns. Nowadays, the use of nanoparticles is widespread and there is a need to regulate nanoparticle manufacturing and utilization for the benefits of individuals, society, as well as ecosystems. Therefore, it is required that policymakers must conclude with some guidelines to control the risks and hazards associated with nanoparticles, as in this modern era, none areas of work are immune from the developments in nanotechnology. Nanotechnology has gained importance in every field of science, society, and humanity in one way or another. This requires that all the members of society must have a communal opinion in its commercialization and further development. This concept proposes a direct interaction between the natural sciences (engineering, technology, mathematics, biology, chemistry, physics, and management) and human development sciences (ethics, psychology, philosophy, sociology, law, and regulations) which govern the morals for nano technological development.

As the development of nanomaterials has increased at an incredible rate in last two decades, the risk or hazard factors associated with them also have increased exponentially.

Recent studies have indicated that lethal nanoparticles exist in nature, i.e. in air, water, soil, and ecosystems, including animals and humans [44, 45]. Toxic nanoparticles are generally found inside or even outside workplaces and may remain suspended in the air, water or soil, which may harm human health directly or indirectly during the course of manufacturing, transportation, distribution, handling, and usage, etc. [45–48]. These suspended nanoparticles may enter the human body through numerous routes including ingestion, skin contact, and oral inhalation, and may cause several adverse effects on human health [49]. The health problems generally associated with the toxic effects of nanoparticles include asthma, respiratory disorders, bronchitis, Parkinson's and Alzheimer's diseases, cancer, and others [50]. The major factor responsible for the greater toxicity of nanoparticles is that they have a larger surface area, i.e. increased site for reactivity and of course different morphology of the surface [51]. A change in the method of producing nanoparticles can also result in variations in surface morphology and hence a different probability of particle reactivity is again generated, due to which humans or the environment may be affected in a different way [52]. The adverse effects of nanoparticles have resulted in the ethics of nano technological research to study for the potential of nanoparticles and analyze their risks and hazards, therefore attracting societal interest, economical concerns, moral and human values, educational culture, law and order formulations, and safety and environmental concerns. There are many fields in which the ethical issues of nanotechnology are associated such as health, society, the environment, biology, and education.

1.4.1 HEALTH ISSUES

This involves the possible effects of nanomaterials on the human body. The most promising ways for nanoparticles to enter the human body are the inhalation, dermal, respiratory, or ingestion routes. Their effect can be variable since different nanoparticles have different surface morphologies and diverse toxicity levels. The nanodevices used for numerous purposes such as information and communication telecast and others can affect human health in two ways: positive health impacts and negative health impacts, as some target-oriented nanoparticles can cure several fatal diseases such as cancer and Alzheimer's disease, whereas some nanoparticles can be the cause of these diseases. For example, short multi-walled carbon nanotubes (MWCNTs) are safer to use as compared to long MWCNTs, as they are less toxic [53].

1.4.2 ENVIRONMENTAL ISSUES

The nano waste that is generated during the processes of manufacturing, transportation, distribution, storage, and waste disposal can be described as nano pollution, which has become a major concern for manufacturing companies as the waste generated in the nano scale should be limited and nontoxic in nature [54]. The nano waste particles generated in a company can migrate to other places or even into the wider environment and, with suitable climatic conditions, they can even penetrate to human or other animal bodies causing greater risks [55]. Several stages of product manufacturing must be assessed from the initial stage of production to the final stage for its toxicity, and a complete life cycle assessment provides all the details about the toxicity level of each stage. This helps in creating a new strategy for the production of goods with low toxicity [56] Sometimes, nano waste production may be useful to some extent, for example, nanoscale membrane or nanoparticles can have the capacity to remove heavy metals from water, hence making it useful [57]. In some studies, it has been shown that nanoparticles can be used to generate energy from fuel cell, solar cells, etc. [58], which could reduce the dependency on nonrenewable resources [59].

1.5   FUTURE SCOPE AND CHALLENGES

The future ramifications of nanobiotechnology are hotly contested. In the areas of medicine, electronics, biomaterials, and alternative energy, they could generate and promote the usage of a variety of new materials and tools. However, this method brings many of the same problems as any new technology, such as concerns about the toxicity and sustainability issues of nanomaterials and their possible impacts on the global economy, as well as speculation about numerous apocalyptic scenarios. Due to these fears, governments and advocacy groups disagree on the need for specific statutory regulation of nanotechnology. Even though there are various concerns, this technology offers great prospects for the future. By playing a major role in a number of biological applications, including in drug administration, gene therapy, molecular imaging, biomarkers, and biosensors, it may result in innovations. Target-specific drugs and techniques for early disease recognition and treatment are two of these applications which are currently the main research focus both in clinical diagnosis and R&D, with two different sorts of therapeutic diagnostics already beginning to emerge [60]. Applications for monitoring cellular processes in material are about to be licensed and imaging applications using nanostructured technology are currently being developed. The creation of extremely precise and sensitive methods of identifying nucleic acids is the second common application type. By 2015 to 2020, products being tested in academic and government laboratories will be creeping into commercialization. Sparse cell isolation and molecular filtration applications should, by then, make it to market. Some of the drug delivery systems should be commercialized or in advanced clinical trials. American Pharmaceutical Partners, investigating the encapsulation of Taxol for cancer treatment using paclitaxel as drug-delivery systems. Most medications and medical devices won't hit the market for at least 10 years. As a result, significant technical infrastructures, such as nanotechnology, as well as highly regulated management are needed for both drug target alteration and device insertion. Nanomedicine has greater potential for use now thanks to current developments in a wide range of medical specialties [61]. Investigations are currently being conducted on its potential use in diagnostic and regenerative medicine. Diagnoses would be made faster, perhaps on the level of a single diseased cell, permitting immediate treatment of sick cells before they spread to and harm other body areas [62]. Nanomedicine may also be useful for people with severe trauma or limited organ functions. Nanobiotechnology has a variety of challenges in promoting its innovations, such as doubts about its efficacy, scalability, finance, a lack of resources, and patience. The large majority of companies acknowledge that nanotechnology has significant promise for the creation of novel products and the enhancement of already-existing ones. Nanotechnology is a novel, potentially disruptive technology that calls into serious question the demand of new laws. Authorities should assess potential risks and the best regulatory methods of action in light of the broad use of this cutting-edge technology.

1.6   CONCLUSION

The key to a sustainable future and its demands are now underlined by nanotechnology and its branches which have attracted the attention of scientists owing to their unique properties and wide range of applications. Scientific applications in various fields have reached new dimensions with the advent of nanotechnology and bionanotechnology. The new realms opened with this technology can be assessed using the advanced instrumentation techniques available but are limited due to the high cost. Although nanotechnology and bionanotechnology have found applications in medical, industrial, agricultural, and environmental domains, they have their own challenges of nano manufacturing and the nano waste generated as a part of the processes involved. Scientists and governments need to work out a way to overcome these challenges and utilize the unique properties of nanoscale materials with advancements in technology and alliances with other branches of science for investigating the unexplored domains.

REFERENCES

[1] Gyorvary, E., Schroedter, A., Talapin, D.V., Weller, H., Pum, D., Sleytr, U.B. (2004) Formation of nanoparticle arrays on S-layer protein lattices. *Journal of Nanoscience and Nanotechnology* 4: 115–120.

[2] Hess, H., Clemmens, J., Brunner, C., Doot, R., Luna, S., Ernst, K.H., Vogel, V. (2005) Molecular self-assembly of "nanowires" and "nanospools" using active transport. *Nano Letters* 5: 629–633.

[3] Montemagno, C. (1999) Constructing nano mechanical devices powered by bio molecular motors. *Journal of Nanotechnology* 10: 225–231.

[4] Liu, H., Schmidt, J.J., Bachand, G.D., et al. (2002) Control of a bio molecular motor-powered Nano device with an engineered chemical switch. *Nature Materials* 1: 173–177.

[5] Xi, J., Schmidt, J.J., Montemagno, C.D. (2005) Self-assembled microdevices driven by muscle. *Nature Materials* 4: 180–184.

[6] Seeman, N.C. (2005) From genes to machines: DNA Nano mechanical devices. *Trends in Biochemical Sciences* 30: 119–125.

[7] Sara, M., Pum, D., Schuster, B., Sleytr, U.B. (2005) S-layers as patterning elements for application in nano biotechnology. *Journal of Nanoscience and Nanotechnology* 5: 1939–1953.

[8] Schuster, B., Gyorvary, E., Pum, D., Sleytr, U.B. (2005) Nanotechnology with S-layer proteins. *Methods in Molecular Biology* 300: 101–123.

[9] Jain, K.K. *Nanobiotechnology in Molecular Diagnosis: Current Techniques and Applications.* Oxford: Taylor & Francis, 2005.

[10] Seeman, N.C. (2006) DNA enables nanoscale control of the structure of matter. *Quarterly Reviews of Biophysics* 6: 1–9.

[11] Niemeyer, C.M., Mirkin, C.A. *Nanobiotechnology: Concepts, Applications and Perspectives.* Weinheim: Wiley-VCH, 2004.

[12] Schutz, C.A., Juillerat-Jeanneret, L., Mueller, H., et al. (2013) Therapeutic nanoparticles in clinics and under clinical evaluation. *Nanomedicine (London)* 8: 449–467.

[13] Kumar, A., Ma, H., Zhang, X., et al. (2012) Gold nanoparticles functionalized with therapeutic and targeted peptides for cancer treatment. *Biomaterials* 33: 1180–1189.

[14] McNeil, S.E. (2011) Unique benefits of nanotechnology to drug delivery and diagnostics. *Methods in Molecular Biology* 697: 3–8.

[15] Caron, J., Reddy, L.H., Lepetre-Mouelhi, S. et al. (2010) Squalenoyl nucleoside monophosphate nanoassemblies: new prodrug strategy for the delivery of nucleotide analogues. *Bioorganic & Medicinal Chemistry Letters* 20: 2761–2764.

[16] Petros, R.A., Desimone, J.M. (2010) Strategies in the design of nanoparticles for therapeutic applications. *Nature Reviews Drug Discovery* 9: 615–627.

[17] Park, W., Na, K. (2015) Advances in the synthesis and application of nanoparticles for drug delivery. *Wiley Interdisciplinary Reviews Nanomedicine and Nanobiotechnology* 7: 494–508.

[18] Elsabahy, M., Wooley, K.L. (2012) Design of polymeric nanoparticles for biomedical delivery applications. *Chemical Society Reviews* 41: 2545–2561.

[19] Wang, Y., Shim, M.S., Levinson, N.S., Sung, H.W., Xia, Y. (2014) Stimuli-responsive materials for controlled release of theranostic agents. *Advanced Functional Materials* 24: 4206–4220.

[20] Yoshida, T., Lai, T.C., Kwon, G.S. (2013) pH- and ion sensitive polymers for drug delivery. *National Library of Medicine* 10(11): 1497–1513.

[21] Nill, K.R. (2007) *Glossary of Biotechnology and Nanobiotechnology Terms*, 4th ed. London: CRC Press 6: 5.

[22] Nalwa, H.S. (2014) A special issue on reviews in biomedical applications of nanomaterials, tissue engineering. *Stem Cells, Bioimaging, and Toxicity* 10: 2421–2423.

[23] Tiwaria, S.K., Bystrzejewski, M., De Adhikari, A., Huczko, A., Wang, N. (2022) Methods for the conversion of biomass waste into value-added carbon nanomaterials: Recent progress and applications. *Progress in Energy and Combustion Science* 92: 101023.

[24] Zeng, M., Gao, H., Yaoqing, W., Aiping, L. (2010) Preparation of bionanomaterials and their polymer nanocomposites from waste and biomass. *Waste and Biomass Valorization* 1: 121–134.

[25] Tosoni S., Chen, H.S.Y.T., Puigdollers, A.R., Pacchioni, G. (2017) TiO_2 and ZrO_2 in biomass conversion: why catalyst reduction helps. *Philosophical Transactions of the Royal Society A* 376: 20170056.

[26] Fortina, P., Kricka, L.J., Surrey, S., Grodzinski, P. (2005) Nanobiotechnology: the promise and reality of new approaches to molecular recognition. *Trends in Biotechnology* 23: 168–173.

[27] Mahmoudi, T., Wang, Y., Hahn, Y.B. (2018) Graphene and its derivatives for solar cells application. *Nano Energy* 47: 51–65.

[28] Kalogirou, S.A. (2004) Solar thermal collectors and applications. *Progress in Energy and Combustion Science* 30: 231–295.

[29] Lee, N.A., Gilligan, G.E., Rochford, J. (2021) Non-innocent ligand flavone and curcumin inspired ruthenium photosensitizers for solar energy conversion. *Physical Chemistry Chemical Physics* 23: 16516–16524.

[30] Lu, J., Chen, Z., Ma, Z., et al. (2016) The role of nanotechnology in the development of battery materials for electric vehicles. *Nature Nanotechnology* 11: 1031–1038.

[31] Ashraf, S.A., Siddiqui, A.J., Abd Elmoneim, O.E., Khan, M.I., Patel, M., Alreshidi, M., Moin, A., Singh, R., Snoussi, M., Adnan, M. (2021) Innovations in nanoscience for the sustainable development of food and agriculture with implications on health and environment. *Science of the Total Environment* 768: 144990.

[32] Malini, S., Kalyan, R., Madhumathy, S., El-Hady, K.M., Islam, S., Dutta, M. (2022) Bioinspired advances in nanomaterials for sustainable agriculture. *Journal of Nanomaterials* 2022: 8926133.

[33] Fraceto, L.F., Grillo, R., De Medeiros, G.A., Scognamiglio, V., Rea, G., Bartolucci, C. (2016) Nanotechnology in agriculture: Which innovation potential does it have? *Frontiers in Environmental Science* 22: 1–5.

[34] Binnig, G., Quate, C.F., Gerber, C. (1986) Atomic force microscope. *Physical Reviews Letters* 56: 930–933.

[35] Geisse, N.A. (2009) AFM and combined optical techniques. *Materials Today* 12: 40–45.

[36] Adrian, R.J. (1991) Particle imaging techniques for experimental fluid mechanics. *Annual Review of Fluid Mechanics* 23: 261–304.

[37] Jesuthasan, N., Baliga, B.R., Savage, S.B. (2006) Use of particle tracking velocimetry for measurements of granular flows: review and application—particle tracking velocimetry for granular flow measurements. *Kona Powder and Particle Journal* 24: 15–26.

[38] Egerton, R.F. (2005) *Physical Principles of Electron Microscopy: An Introduction to TEM, SEM, and AEM*. Springer, New York.

[39] Carlton, C.E., Ferreira, P.J. (2012) In situ TEM nanoindentation of nanoparticles. *Micron* 43: 1134–1139.

[40] Cumberland, S.L., Hanif, K.M., Javier, A., Khitrov, G.A., Strouse, G.F., Woessner, S.M., Yun, C.S. (2002) Inorganic clusters as single-source precursors for preparation of CdSe, ZnSe, and CdSe/ZnS nanomaterials. *Chemical Materials* 14: 1576.

[41] Lin, H. (2006) Datar RH: Medical applications of nanotechnology. *The National Medical Journal of India* 19: 27–32.

[42] Guccione, S., Li, K.C., Bednarski, M.D. (2004) Vascular-targeted nanoparticles for molecular imaging and therapy. *Methods in Enzymology* 386: 219–236.

[43] Farjadian, F., Ghasemi, A., Gohari o Roointan, A., Karimi, M., Hamblin, M.R. (2019) Nanopharmaceuticals and nanomedicines currently on the market: challenges and opportunities. *Nanomedicine London* 1: 93–126.

[44] Kumar, C. (2006) *Nanomaterials: Toxicity, Health and Environmental Issues*. Wiley-VCH.

[45] Karakoti, A.S., Hench, L.L., Seal, S. (2006) The potential toxicity of nanomaterials: The role of surfaces. *Journal of the Minerals, Metals and Materials Society* 58: 77–82.

[46] Brayner, R. (2008) The toxicological impact of nanoparticles. *Nanotoday* 3: 48–55.

[47] Maynard, A.D. (2006) Safe handling of nanotechnology. *Nature* 444: 267–269.

[48] Asmatulu, R., Khan, W.S., Nguyen, K.D., Yildirim, M.B. (2010) Synthesizing magnetic nanocomposite fibers for undergraduate nanotechnology education. *International Journal of Mechanical Engineering Education* 38: 196–203.

[49] Asmatulu, R., Asmatulu, E. (2011) Importance of recycling education: A curriculum development at Wichita State University. *Journal of Material Cycles and Waste Management* 134: 131–138.

[50] Department of Health and Human Services (2009) *Centers for Disease Control and Prevention National Institute For Occupational Safety and Health, Approaches to Safe Nanotechnology*. DHHS (NIOSH) Publication No. 03: 125.

[51] Asmatulu, R., Zhang, B., Asmatulu, E. (2013) *Safety and Ethics of Nanotechnology* 10: 16. doi.10.1016/B978-0-444-59438-9.00003-5

[52] Sandler, R. (2009) *Nanotechnology: The Social and Ethical Issues*. The Pew Charitable Trust.

[53] Olmstead, M., Bassett, D. (2009) *Teaching Nanoethics to Graduate Students, Special Edition Monograph: Nanoethics Graduate Education Symposium*. University of Washington.

[54] U.S. Environmental Protection Agency (2007) www.epc.gov.

[55] Nuraje, N., Asmatulu, R., Kudaibergenov, S. (2012) Metal oxide-based functional materials for solar energy conversion: A review. *Current Inorganic Chemistry* 2: 124–146.

[56] Nuraje, N., Khan, W.S., Ceylan, M., Lie, Y., Asmatulu, R. (2013) Superhydrophobic electrospun nanofibers. *Journal of Materials Chemistry A* 1: 1929–1946.

[57] Asmatulu, R., Asmatulu, E., Zhang, B. (2012) Recent progress in nanoethics and its possible effects on engineering education. *International Journal of Mechanical Engineering Education* 40: 1–10.

[58] Buzea, C., Pacheco, I., Robbie, K. (2007) Nanomaterials and nanoparticles: Sources and toxicity. *Biointerphases* 2: MR17.

[59] Sahoo, K.S., Labhasetwar, V. (2003) Nanotech approaches to drug delivery and imaging. *Drug Discovery Today* 8(24): 1112–1120.

[60] Hamad-Schifferli, K., Schwartz, J., Santos, A.T., Zhang, S., Jacobson, J. (2002) Remote electronic control of DNA hybridization through inductive coupling to an attached metal nanocrystal antenna. *Nature* 415: 152–155.

[61] Curtis, A., Wilkinson, C. (2001) Nanotechniques and approaches in biotechnology. *Trends in Biotechnology* 19: 97–101.

[62] Sastry, R.K., Rashmi, H., Rao, N. (2011) Nanotechnology for enhancing food security in India. *Food Policy* 36: 391–400.

2 Biological Self-Assembly in Bionanotechnology

Sarabjot Kaur Makkad

2.1 INTRODUCTION

From molecules to galaxies, self-assembly is a ubiquitous phenomenon that governs the structural organization at all scales. It can be described as automatic processes of organization where individual components are organized and assembled to highly ordered patterns without any external intervention. According to Tecilla et al., self-assembly can be described as a "non-covalent interaction of two or more molecular subunits to form an aggregate whose novel structure and properties are determined by nature and positioning of individual components" [1]. Whitesides et al. defined self-assembly as "spontaneous association and organization of individual components to form more structured, stable and non-covalently joined architectures" [2,3].

When the individual components are molecules, then such self-assembly is molecular self-assembly. It can be elucidated as "spontaneous organization of molecules under near thermodynamic equilibrium conditions into structurally well-defined and stable arrangements through non-covalent interactions" [4]. Such self-assembly is primarily governed by weak non-covalent interactions such as hydrogen bonding, van der Waal's interaction, hydrophobic and hydrophilic interactions, and electrostatic interactions [4,5]. Although these non-covalent interactions are weak, collectively, due to chemical complementarity and structural compatibility, these weak forces results in chemical and structural stability [4,6].

Self-assembly in a biological system is particularly important and can be evidently observed in the formation of various supramolecular nanostructures such as DNA, RNA, peptides, proteins, and phospholipid bilayers of cell membrane. The formations of bacteriophages and virus particles are also based on the principle of self-assembly (Figure 2.1).

Self-assembly in nanotechnology has been described as the "creation of material from its constituent components in spontaneous 'natural' manner, i.e. by an interaction between the components or by specific rearrangement of them, that proceeds naturally without any special external impetus" [7]. Bionanotechnology is the extension of nanotechnology that utilizes biology for technological applications at the nanoscale. In order to maximize the innate properties of biomolecules, the "bottom-up" assembly technique is employed in bionanotechnology [8].

DOI: 10.1201/9781003362258-2

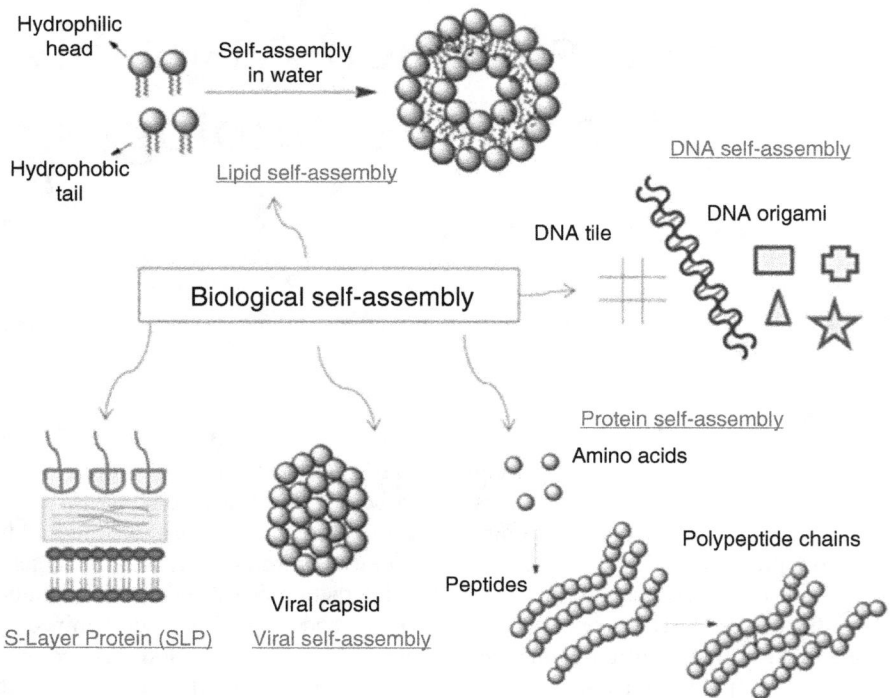

FIGURE 2.1 Schematics of biological self-assembly.

2.2 CLASSIFICATION OF BIOLOGICAL SELF-ASSEMBLY

Self-assembly on the basis of size and nature of building blocks can be broadly classified into three categories, i.e. atomic, molecular, and colloidal self-assembly.

Based on systems, it is categorized into *biological* and *interfacial* self-assembly. It is classified into *thermodynamic* and *kinetic* self-assembly based on processing. Atomic, molecular, biological, and interfacial self-assembly fall under the category of thermodynamic processes, while colloidal and some interfacial self-assembly fall under kinetic processes. Additionally, biological and atomic self-assembly are directional, while the remaining self-assemblies including interfacial, colloidal, and molecular are non-directional.

Based on energy dissipation, self-assembly can be classified into two types: *static* and *dynamic* [9]. In static self-assembly, energy is required only during the formation of the highly ordered assembled structure but once it is formed, it does not require further energy and is stable. For instance, haemoglobin polypeptide forming functional haemoglobin protein, oil droplets from lipid molecules in water, ribosomal protein and RNA forming functional ribosome are some significant examples of static self-assembly. In dynamic self-assembly, the system requires continuous energy dissipation for each interaction taking place between the components for the formation of patterns or structures [9,10]. The system will disassemble as soon as the energy ceases to flow into it. Examples of dynamic self-assembly include all living organisms that thrive on energy available from food, and these organisms would disassemble as soon as the energy flux is ceased.

The process of static and dynamic self-assembly can be further sub-classified into three categories: co-assembly, directed self-assembly, and hierarchical self-assembly [10]. In *co-assembly*, the different constituents are self-assembled within the same system in a synergetic manner. In *hierarchical* self-assembly, constituents of the same kind self-assemble to form a first-order assembly which in turn serves as a building block for second-order assembly and so on. This type of self-assembly often leads to several orders of assembly. Biological systems employ this type of assembly to form huge functional structures. Assembly is directed in the case of *directed* self-assembly by external forces.

2.3 LIPID-BASED SELF-ASSEMBLY

Lipids are hydrocarbon-based biomolecules that are soluble in nonpolar solvent and insoluble in polar solvent. Oils, wax, fats, and vitamins are some common examples of lipids. Amphiphilic lipid-based self-assembly leads to various nanostructures including micelles, vesicles, liposomes, nanotubes, and nanofibers. Such self-assembled nanostructures have found considerable applications in the field of science and technology. An individual phospholipid has an approximate dimension of 2.5 nm. It consists of a phosphate head as the hydrophilic head and an aliphatic hydrophobic tail. Being amphiphilic, it assembles spontaneously in water via self-assembly. Phosphatidic acid, phosphatidylglycerol, phosphatidylserine, and phosphatidyl ethanolamine are some commonly used natural anionic phospholipids. Stearylamine that is used to encapsulate nucleic acid is an example of natural cationic phospholipid.

The thermodynamic parameter of self-assembly has been determined using isothermal titration calorimetry by Marsh et al. [11]. Large heat capacity and zero enthalpy of dioctanoyl phosphatidylglycerol and phosphatidylserine at particular temperatures having a hydrophobic effect (attributed by aliphatic lipid chains) are the main driving forces for self-assembly.

In a phospholipid bilayer structure, aliphatic chains in each layer are pointed toward each other, while hydrophilic phosphate heads are pointed outside toward water to form a close structure (Figure 2.2). This self-assembly results in nanostructures such as nanospheres, nanotubes, etc. According to the literature, most common amphiphiles have a critical micellar concentration (CMC) of 10^{-2} to 10^{-4} M, while for phospholipids, it is even four to five times less. This simply implies that they are more stable than micelles after administration [12]. Phosphatidic acid, phosphatidylglycerol, phosphatidylserine, and phosphatidyl ethanolamine are some of the commonly used natural anionic phospholipids. Stearylamine that is used to encapsulate nucleic acid is an example of a natural cationic phospholipid. Natural phospholipids exhibit higher chemical biostability against many common enzymes and salts such as phospholipase, bile salt, esterase, and serum proteins present in the body. Therefore, vesicles formed from natural phospholipids are thermodynamically more stable in terms of alkaline pH, oxidative stress, and high temperature as compared

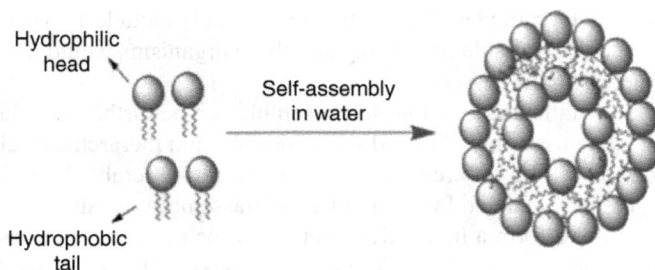

FIGURE 2.2   Schematics of lipid-based self-assembly.

to synthetic phospholipids. The next frequently used lipid in nanoengineered drug-delivery systems is cholesterol [12]. Kirby et al. showed that cholesterol comprising liposomal membranes is more stable as compared to membrane lacking cholesterol in blood [12]. X-ray-triggered [13] and photo-triggered [14] liposomal-based systems have been reported for sustained drug release. However, there is plenty of room to further develop liposomal technology to be used for drug-delivery applications [15,16].

Liposomes are amphiphilic lipids that self-assemble in water spontaneously to form hollow spherical vesicles (Figure 2.2). Liposomes are the most widely used bionanomaterials in cancer therapy. They are used to encapsulate bioimaging agents, enzymes, and drug molecules to the target site. Due to their vesicular nature, liposomes can encapsulate both hydrophobic and hydrophilic drugs simultaneously. Additionally, liposomes are characterized by their non-toxicity and biodegradability.

The first liposome-based anticancer formulation which was approved by the FDA in 1995 for AIDS-associated Kaposi's sarcoma was Doxil [17].

The anticancer drug encapsulated inside Doxil is doxorubicin. Doxorubicin is an anthracycline which exerts anticancer activity by arresting DNA and RNA synthesis. However, it causes severe myelosuppression and cardiotoxicity [12] as side effects, which can be even more severe than the disease itself. These side effects could be diminished to some extent by encapsulating doxorubicin in a liposome [18].

The drug was loaded inside liposomes by an active drug-loading technique for enhancing drug-loading content and stability, and this resulted in two drug formulations, i.e. Myocet and Doxil. In Doxil, the potential gradient was used as a trigger for drug loading, while in Myocet, the pH gradient was used for encapsulating drug inside liposomes. PEG coating in Doxil enhanced its pharmacokinetics as compared to Myocet. Both these formulations proved to be successful in reducing the side effects of doxorubicin [19].

Other liposome-based drugs include Ambisome which encapsulates amphotericin, an antifungal drug, Depodur which encapsulates morphine, and Visudyne which encapsulates verteporfin (used for the treatment of macular degeneration).

Additionally, liposomes also find application in the cosmetic industry to encapsulate vitamin E, vitamin C, and other antioxidants. For example, Laouini et al. encapsulated vitamin E with a high efficiency of 99.87% inside size-controlled liposomes using a new approach [20]. The size of the liposomes was maintained using stearic acid or cholesterol. Such a formulation was proved to be effective in the treatment of lungs. Moreover, Ismail et al. synthesized dimeric artesunate phospholipid-based novel liposomes having multilamellar vesicles with bilayer morphology [21]. This formulation exhibited antiplasmodial activity.

2.4 PEPTIDE- AND PROTEIN-BASED SELF-ASSEMBLY

Proteins are complex macromolecules composed of repeating units of amino acids. Amino acids contain one amino group ($-NH_2$) and one carboxyl group (-COOH). Two or more amino acids condense to form a peptide molecule via a peptide bond. Many peptide molecules align themselves in a particular pattern to form different types of proteins. For instance, antibodies, enzymes, and hormones are some examples of proteins in biological systems. Nature has wonderfully utilized the self-assembly of proteins and peptides to create a wide range of structures such as pearl, coral, shell, collagen, keratin, etc. The self-assemblies of proteins and peptides have found immense applications in the fields of biomedicine and bionanotechnology and are successfully utilized in the synthesis of bionanomaterials [6,8,22,23].

Peptides are short amino acid chains as compared to proteins and are usually comprised of 2–50 amino acids [24]. Peptides are biocompatible and biodegradable molecules, which makes them suitable for biomedical applications. Additionally, due to their shorter chain length, peptides are easy to design and synthesize. This class of bionanomaterials is versatile and used in a large number of applications including drug delivery, biological surface engineering, and tissue repair and engineering.

Proteins and peptides, like lipids, can also self-assemble to vesicles, micelles, and nanotubes (Figure 2.3). One can theoretically predict the nature of nano-assembled structures using critical packing factors (Cpp):

$$Cpp = V/Al$$

> where V is hydrophobic volume
> A is hydrophilic head area
> l is hydrophobic chain length.

Type I peptides, also known as "molecular Lego," are the class of peptides having an amphiphilic structure that can self-assemble into beta sheets in water. They have the unique ability to form self-complementary ionic bonds with alternative hydrophilic and hydrophobic residues. Moreover, in response to a change in electrolyte concentration or pH, these stable beta sheet structures result in nanofibers. Many such nanofibers are interwoven and self-assembled together to give a scaffold hydrogel having high water content. Tirrell et al. [26] were the first to report hydrogel formed from self-assembly of artificial proteins in response to a change in pH and temperature. These hydrogels find applications in advanced wound healing and tissue repair [23,27]. Such a self-assembly design could easily be extended to polymers and their composites.

Type II, also called "molecular switches," are the class of peptides that undergo drastic conformational transitions under certain environmental conditions such as a

C_{pp}	Geometry amphiphilic molecule	Structure
$p<\dfrac{1}{3}$	Cone	
$\dfrac{1}{3}<p<\dfrac{1}{2}$	Truncated cone	
$\dfrac{1}{2}<p<1$	Truncated cone	
$p>1$	Inverted truncated cone	

FIGURE 2.3 Effect of critical packaging factor (Cpp) on the geometry of amphiphilic peptides.

Source: Adapted from Ref. [25].

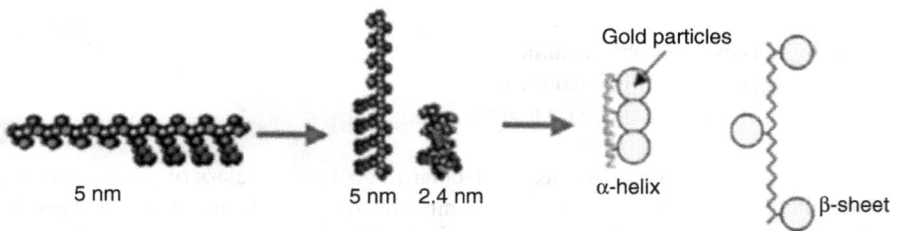

Gold particles

5 nm   5 nm 2.4 nm α-helix β-sheet

FIGURE 2.4 Schematics of molecular switch undergoing drastic conformational change.

Source: Adapted from Ref. [30].

change in temperature, pH, location, or crystal lattice packing [28]. Additionally, like Type I, this class also has complementary ionic bonds; however, it is more dynamic and complicated as compared to Type I. For instance, at room temperature, DAR16-IV has a normal beta sheet structure with 5 nm length (Figure 2.4) but at higher temperature, it undergoes drastic structural change from a beta sheet to a stable alpha helical structure of 2.5 nm length [29,30]. On self-assembly, such peptides result in different structures depending on the size and shape of the polar hydrophilic head. For example, V6D peptide can be self-assembled into nanovesicles or nanotubes in water. However, such self-assembled structures can be altered by changing environmental factors. Self-assembled membranes from these peptide nanotubes can be used in nanobiosensor devices [31]

Type III is often called "molecular paint" or "molecular Velcro." Unlike Types I and II, Type III has no self-complementarity and would rather self-assemble onto the surface. These peptides engineer surfaces with three features, i.e. surface anchor, central linker, and terminal segment of ligand. This technique could provide a new method for cell–cell interactions and studying their behaviour. Moreover, some amphiphilic peptides can self-assemble to nanotubes and nanovesicles, and are included in Type IV. However, their broader size distribution over time indicates the extremely dynamic process of assembly and disassembly.

Thus it is evident that peptides and proteins can self-assemble into various nanostructures such as nanovesicles, nanotubes, nanofibers, etc. These nanostructures have an immense range of applications in the fields of bionanotechnology, tissue engineering, regenerative medicine, and dentistry. Therefore, their self-assembly will pave new ways for the development of novel bionanomaterials.

2.5   DNA-BASED SELF-ASSEMBLY

DNA is a robust tool for many nanostructures and shapes. It is a polynucleotide that stores as well as transmits genetic information and forms the genetic basis of life. It consist of two main nucleobases, namely purines (adenine and guanine) and pyrimidines (cytosine, uracil, and thymine). DNA as a nanofabrication template is a leading edge tool for developing customized nanomaterials using its self-assembly phenomenon. The fundamental basis of DNA bionanotechnology is complementary base pairing.

The control and synthesis of more complicated nanostructures could be achieved using two main self-assembly techniques: (i) DNA origami and (ii) single-stranded tiles (SST).

(i)   DNA origami:
       Folding of long single-stranded DNA helices at the nanoscale is used to create specified 2D and 3D shapes. This can be achieved by utilizing various short oligonucleotides called staple strands to instigate folding of long single-stranded DNA (scaffold). The scaffold DNA is mixed with staple strands followed by heating and annealing. The primary prerequisite for DNA origami is a long DNA strand which is acquired from M13 bacteriophage. In 2005, Rothemund [32] coined the term "DNA origami" and developed a range of 2D nanostructures, namely triangle, star, rectangle, and smiley (Figure 2.5). However, these 2D DNA origami can be folded into 3D DNA origami by interconnecting staple strands. 3D DNA origami can also be designed by packaging DNA helix and its positioning by DNA crossovers. Various DNA

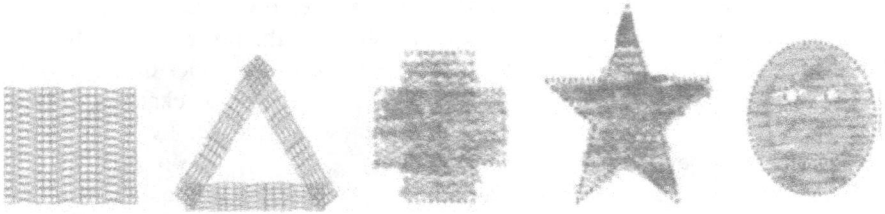

FIGURE 2.5   Schematics of various DNA origami nanostructures.

Source: Adapted from Ref. [34].

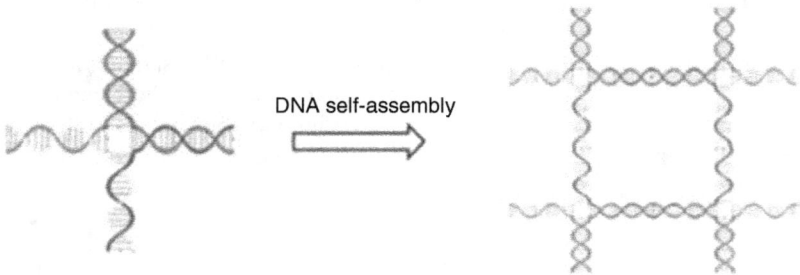

DNA self-assembly

FIGURE 2.6   Schematics of DNA self-assembly into tile.

Source: Adapted from Ref. [34].

origami nanostructures were reported by Du et al. for doxorubicin delivery [33]. Doxorubicin, being an efficient DNA intercalating molecule, could be loaded to DNA origami nanostructures. These nanostructures showed an enhanced drug-loading efficiency as well as better cellular uptake [33]. The hierarchical scaling up of DNA origami includes two strategies, i.e. sticking end base pairing and blunt end base stacking.

(ii) Single-stranded tile (SST):

Single short-stranded tiles are formed from several short single-stranded DNA molecules (Figure 2.6). Various nanostructure motifs can be constructed by linking sticky ends of DNA strands. Unpaired nucleotide overhangs present on the sticky ends combine together with the complementary strand by hybridization. DNA motifs are repeatedly arranged to form 2D structures by sticky end base pairing. High-order DNA motifs can be created using a simple DX structure. The basic rule for designing DNA tile nanostructures implies the minimization of sequence symmetry in the branch structure to circumvent any chances of undesired pairing among the DNA strands [35]. However, when compared to DNA origami, it has poor size and shape control of product. Table 2.1 presents a comparison between DNA tile and DNA origami nanostructures.

TABLE 2.1
Comparison between DNA Tile and DNA Origami

Self-assembly method	DNA Tiles	DNA Origami
Starting material	Several short single-stranded oligonucleotides	Hundreds of short single-stranded oligonucleotides and long oligonucleotides
Scaling-up strategies	Sticky end base pairing	Sticky and blunt end base pairing
Type of assembly	2D or 3D framework	1D, 2D, or 3D framework
Assembly dimension	Up to 1 mm	10 μm

2.6 S-LAYER PROTEIN SELF-ASSEMBLY

The outer surfaces of both archaea and bacteria are coated with 2D crystalline arrays of proteinaceous S-layer. Crystalline lattices consisting of single glycoprotein are self-assembled into a regular 2D array on the surface of cells. When compared to Gram-negatives, Gram-positives have more S-layer-associated gene families. It is often considered as an evolutionary adaptation to particular ecological and environmental conditions by the organism. S-layer glycoproteins are the most abundantly expressed proteins and have been recognized in many species of almost every group of walled bacteria. Additionally, it signifies the simplest yet most abundant biological protein on the planet. According to studies, it can serve as promoters for cell adhesion, surface recognition, and anti-fouling coating, and also function as a virulence factor in pathogenic organisms, protective coats, and molecular sieves. S-layer proteins have played a crucial role in self-assembly research mainly because of ability of SLPs to assemble themselves into crystalline and ordered nanomaterials. A 2D array consists of two main domains: (i) an anchoring domain that anchors S-layer protein to the cell wall and (ii) a crystalline domain that is responsible for inter-S-layer protein interactions. They displayed similar physiochemical properties at the nanoscale and have openings that are identical in size and morphology in the size range of 2–8 nm. Moreover, properties of SLPs can be tuned by genetic engineering and modification.

There are three major classes of SLP profile [36] (Figure 2.7): (i) in Archaea, S-layer glycoproteins mostly represent either (a) a mushroom-like structure with pillar domain or (b) a lipid-modified protein structure. Meanwhile, a few archaea possess (c) a rigid layer in between the S-layer and plasma membrane. (ii) SLPs are adhered to a peptidoglycan layer of the cell wall in Gram-positive bacteria. (iii) Peptidoglycan is much thinner in the case of Gram-negative bacteria, therefore SLPs are attached to lipopolysaccharides, which are on the outer layer of cell membrane. One can identify the S-layer of any organism by freeze etching of pristine cells using TEM. 2D crystalline organization can be identified using electron crystallography [37], scanning probe microscopy [38], X-ray, or neutron-scattering techniques [37]. Lattices of S-layers displayed square, oblique, and hexagonal symmetry [37]. The thickness of bacterial S-layers is approximately 5–10 nm, while it is much thicker in archaea. Nanobiotechnology applications based on the S-layer mainly rely on the ability of

A Gram-negative archaea

S-layer

Cytoplasmic
membrane

B Gram-positive bacteria and archaea

S-layer

Peptidoglycan or
other polymers

Cytoplasmic
membrane

C Gram-negative bacteria

S-layer

Outer membrane

Peptidoglycan

Cytoplasmic
membrane

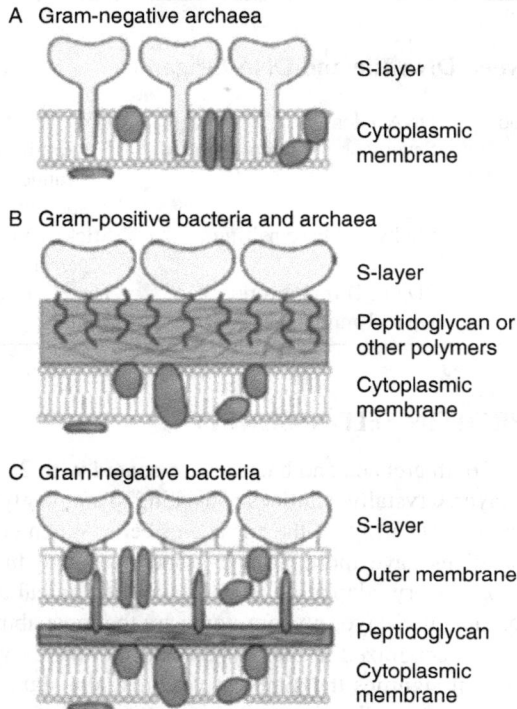

FIGURE 2.7 Schematic illustration of major classes of prokaryotic cell containing surface
(S) protein.
Source: Adapted from Ref. [36].

individual S-layer proteins to self-assemble in vitro. In vitro self-assembly is directed
by the amino acid sequence of protein chains. As SLPs contain a higher proportion of
non-polar amino acids, in vitro assembly is mainly based on hydrophobic interactions.
In vitro self-assembly can be observed in both suspensions and solid supports, such as
silicon, metal, polymers, and lipid membrane. Additionally, SLP-incorporated functional domains, e.g. antigens, antibodies, enzymes, fluorescent proteins, etc. provide
a broad spectrum of applications in areas such as diagnostics, vaccines, microcarriers,
and other biomimetic applications [39–41]. Moreover, SLP-based technologies have
immense possibilities for biological templating, nanopatterning of surfaces, and
microfluidic devices.

2.7 VIRUS-BASED SELF-ASSEMBLY

Being the most abundant biological agents on the planet, viruses play a significant role
in biological systems. Tobacco mosaic virus (TMV) is one of the best studied viral
self-assembly systems. It is also an example of the first macromolecular nanostructure

to be self-assembled in vitro. It is a rod-shaped plant virus with a length of 3000Å and diameter of 180Å. Its capsid consists of (i) capsomer (protein shell of approximately 2130 identical protein subunits) that surrounds (ii) single-stranded RNA as its genetic material which is 6395 nucleotides long. Capsomer can self-assemble into a rod-shaped right-handed helical assembly with 16.3 proteins per helix turn and an inner channel of 40Å. However, this protein shell has the ability to undergo shape transformation from rods in pH 5.5 (acetate buffer) to double disk in pH 7.0 (phosphate buffer). Both external as well as internal surfaces of capsids are comprised of regular patterns of charged amino acids, namely aspartate, glutamate, arginine, and lysine that are available for bioconjugation and chemical ligation [42]. TMV virus can be isolated in kilogram quantities from an infected tobacco plant. Additionally, this virus has the ability to self-assemble in vitro and can withstand temperatures of up to 80°C and pH between 2–10. The genetic material, i.e. RNA, is loaded into the capsid in vivo via an assembly and disassembly process. Under proper conditions of pH and ionic strength, this capsomer can assemble spontaneously into non-infectious viral-like particles known as organic nanoparticles. This process of encapsulation of RNA into capsid can be utilized to load any functional moiety namely enzymes, polymers, liquid droplets, genes, etc. [43]. It can be achieved by simply mixing protein subunits with functional moiety at specific temperature, pH, and ionic strength. In this respect, several explorations have been carried out using TMV and other viruses such as CCMV, brome mosaic virus (BMV), bacteriophage MS2, JC virus, etc.

Thus, all such viruses and a variety of virus-like particles finds applications such as imaging agents, gene- and drug-delivery vehicles, and reaction vessels. It is a potential candidate for drug delivery due to its biocompatibility and biodegradability. By altering types of RNAi and SiRNA, many diseases from genetic disorders to cancer could be targeted. Bentley and team [44] employed hollow TMV capsid to load RNAi for a gene-delivery application. In RNA interference (RNAi), the RNA molecule is used to reduce the gene expression of specific mRNA. Ijiro et al. [45] developed a pH-dependent drug-release system based on JC virus, a human polyomavirus. Green fluorescent protein (GFP) was encapsulated inside capsid and its controlled release was further studied. Designing of a cell-specific drug vehicle could be achieved by tuning the ease of encapsulation and release of drug.

The virus capsid can also act as an enzyme nanocontainer by encapsulating enzyme in a specific manner to have catalytic activity. *Pseudozyma antarctica* lipase B (PalB) enzyme loaded into CCMV capsid resulted in an evident overall increase in the rate of reaction when compared to non-encapsulated capsid [46].

Near-infrared (NIR) imaging is particularly important as it prevents autofluorescence from cells. Indocyanine green, the only FDA-approved NIR dye. is loaded into BMV capsid to form optical viral ghosts (OVGs) with a mean diameter of 24.3 nm. It was observed that after 3 h, more than 90% of the OVGs were internalized into human bronchial epithelial cells. Such constructs have the potential for deep tissue imaging and disease therapy [47]. TMV with a free -SH group can also be utilized as a light-harvesting system through thiol reactive attachment of the chromophore [48]. For instance, donor and acceptor chromophores can be installed on the inner side of TMV such that there is efficient energy transfer from the donor to the acceptor [48]. This has emerged as a prominent technology for photovoltaic applications. The capability of virus capsid to encapsulate a wide variety of functional moiety has resulted in a range of novel functional nanomaterials. Cu (I) catalysed azide alkyne cycloaddition (CuAAC) and diazonium coupling could be employed for the attachment of virus to the functional moiety [49]. These unique architectures will have applications in areas such as drug delivery, enzyme confinement, biosensing, light harvesting systems, etc.

REFERENCES

1. Tecilla, P., Dixon, R.P., Slobodkin, G., Alavi, D.S., Waldeck, D.H., Hamilton, A.D. (1990) Hydrogen-bonding self-assembly of multichromophore structures. *Journal of the American Chemical Society* 112: 9408–9410.
2. Whitesides, G.M., Mathias, J.P., Seto, C. (1991) Molecular self-assembly and nanochemistry: a chemical strategy for the synthesis of nanostructures. *Science* 254: 1312–1319.
3. Whitesides, G.M., Boncheva, M. (2002) Beyond molecules: self-assembly of mesoscopic and macroscopic components. *Proceedings of the National Academy of Sciences USA* 99: 4769.
4. Zhang, S. (2001) *Molecular Self-Assembly, Encyclopedia of Materials: Science & Technology*, p. 5822. Elsevier Science, Oxford, UK.
5. Pelesko, J.A. (2007) *Self Assembly: The Science of Things That Put Themselves Together*, p. 5. Chapman & Hall/CRC, London.
6. Zhang, S. (2003) Fabrication of novel biomaterials through molecular self-assembly. *Nature Biotechnology* 21: 1171.
7. http://nanoatlas.ifs.hr/index_2.html.
8. Goodsell, D.S. (2004) *Bionanotechnology: Lessons from Nature*. Wiley-Liss.
9. Whitesides, G.M., Grzybowski, B. (2002) Self-assembly at all scales. *Science* 295: 2418.
10. Cademartiri, L., Ozin, G.A. (2009) *Concepts of Nanochemistry*, Wiley-VCH.
11. Marsh, D. (2012) Thermodynamics of phospholipid self-assembly. *Biophysical Journal* 102: 1079–1087.
12. Felice, B., Prabhakaran, M.P., Rodriguez, A. P., Ramakrishna, S. (2014) Drug delivery vehicles on a nano-engineering perspective. *Materials Science and Engineering C* 41: 178–195.
13. Deng, W., Chen, W., Clement, S., Guller, A., Zhao, Z., Engel, A., et al. (2018) Controlled gene and drug release from a liposomal delivery platform triggered by X-ray radiation. *Nature Communications* 9: 2713.
14. Massiot, J., Rosilio, V., Makky, A. (2019) Photo-triggerable liposomal drug delivery systems: from simple porphyrin insertion in the lipid bilayer towards supramolecular assemblies of lipid–porphyrin conjugates. *Journal of Materials Chemistry B* 7: 1805–1823.
15. Pattni, B.S., Chupin, V.V., Torchilin, V.P. (2015) New developments in liposomal drug delivery. *Chemical Reviews* 115: 10938–10966.
16. Salunkhe, S.S. et al. (2013) Liposomes: Benchmark in the era of drug carriers for semisolid based topical delivery system. *American Journal of PharmTech Research* 3: 494–510.
17. Northfelt, D.W., Dezube, B.J., Thommes, J.A., Miller, B.J., Fischl, M.A., Friedman-Kien, A., et al. (1998) Pegylated-liposomal doxorubicin versus doxorubicin,

bleomycin, and vincristine in the treatment of AIDS-related Kaposi's sarcoma: results of a randomized phase III clinical trial. *Journal of Clinical Oncology* 16: 2445–2451.

18. Fan, Y., Zhang, Q. (2013) Development of liposomal formulations: from concept to clinical investigations. *Asian Journal of Pharmaceutical Sciences* 8: 81–87.

19. Hofheinz, R.-D., Gnad-Vogt, S. U., Beyer, U., Hochhaus, A. (2005) Liposomal encapsulated anti-cancer drugs. *Anticancer Drugs* 16: 691–707.

20. Laouinia, A., Charcosset, C., Fessi, H., Holdich, R.G., Vladisavljević, G.T. (2013) Preparation of liposomes: a novel application of microengineered membranes – investigation of the process parameters and application to the encapsulation of vitamin E. *RSC Advances* 3: 4985–4994.

21. Ismail, M., Ling, L., Du, Y., Yao, C., Li, X. (2018) Liposomes of dimeric artesunate phospholipid: a combination of dimerization and self-assembly to combat malaria. *Biomaterials* 163: 76–87.

22. Segers, V.F.M., Lee, R.T. (2007) Local delivery of proteins and the use of self-assembling peptides. *Drug Discovery Today* 12: 561.

23. Halstenberg, S., Panitch, A., Rizzi, S., Hall, H., Hubbell, J.A. (2002) Biologically engineered protein-graft-poly(ethylene glycol) hydrogels: A cell adhesive and plasmin-degradable biosynthetic material for tissue repair, *Biomacromolecules* 3: 710.

24. Lintner, K., Peschard, O. (2000) Biologically active peptides: from a laboratory bench curiosity to a functional skin care product. *International Journal of Cosmetic Science* 22(3): 207–218.

25. Lombardo, D., Kiselev, M.A., Magazu, S., Calandra, P. (2015) Amphiphiles self-assembly: Basic concepts and future perspectives of supramolecular approaches. *Advances in Condensed Matter Physics* 2015: 151683.

26. Petka, W.A., Harden, J.L., McGrath, K.P., Wirtz, D., Tirrell, D.A. (1998) Reversible hydrogels from self-assembling artificial proteins. *Science* 281: 38.

27. Vauthey, S., Santoso, S., Gong, H., Watson, N., Zhang, S. (2002) Molecular self-assembly of surfactant-like peptides to form nanotubes and nanovesicles. *Proceedings of the National Academy of Sciences USA* 99: 5355.

28. Altman, M., Lee, P., Rich, A., Zhang, S. (2000) Conformational behavior of ionic self-complementary peptides. *Protein Science* 9: 1095.

29. Zhang, S., Marini, D.M., Hwang, W., Santoso. S. (2002) Design of nano biological materials through self-assembly of peptides and proteins. *Current Opinion in Chemical Biology* 6: 865.

30. Subramani, K., Ahmed. W. (2012) Self-assembly of proteins and peptides and their applications in bionanotechnology and dentistry. *Emerging Nanotechnologies in Dentistry* 209–224.

31. Vauthey, S., Santoso, S., Gong, H., Watson, N., Zhang, S. (2002) Molecular self-assembly of surfactant-like peptides to form nanotubes and nanovesicles. *Proceedings of the National Academy of Sciences USA* 99: 5355.

32. Rothemund, P.W.K. (2006) Folding DNA to create nanoscale shapes and patterns. *Nature* 440: 297–293.

33. Zhang, Q., Jiang, Q., Li, N., Dai, L., Liu, Q., Song, L., Wang, J., Li, Y., Tian, J., Ding, B., Du, Y. (2014) DNA origami as an in vivo drug delivery vehicle for cancer therapy. *ACS Nano* 8: 6633–6643.

34. Sharma, A., Vaghasiya, K., Verma, R. K., Yadav, A. B. (2018) DNA nanostructures: Chemistry, self-assembly, and applications. In: *Emerging Applications of Nanoparticles and Architectural Nanostructures: Current Prospects and Future Trends*, pp. 71–94. Elsevier Inc.

35. Lin, C., Liu, Y., Rinker, S., Yan, H. (2006) DNA tile based self-assembly: building complex nanoarchitectures. *ChemPhysChem* 7: 1641–1647.
36. Sleytr, U.B., Schuster, B., Egelseer, E.M., Pum, D., Horejs, C.M., Tscheliessnig, R., Ilk, N. (2011) Nanobiotechnology with S-layer protein as building blocks. *Progress in Molecular Biology and Translational Science.* 103: 277–333.
37. Pavkov-Keller, T., Howorka, S., Keller, W. (2011) The structure of bacterial S-layer proteins. In: *Molecular Assembly in Natural and Engineered Systems*, Vol. 103 (Howorka S, ed.), pp. 73–130. Elsevier Academic Press Inc., Burlington.
38. Lopez, A.E., Pum, D., Sleytr, U.B., Toca-Herrera, J.L. (2011) Influence of surface chemistry and protein concentration on the adsorption rate and S-layer crystal formation. *Physical Chemistry Chemical Physics* 13: 11905–11913.
39. Egelseer, E.M., Ilk, N., Pum, D., Messner, P., Schaffer, C., Schuster, B., Sleytr, U.B. (2010) S-layers, microbial, biotechnological applications. In: Flickinger MC, editor. *The Encyclopedia of Industrial Biotechnology: Bioprocess, Bioseparation, and Cell Technology*, Vol. 7. pp. 4424–4448. Hoboken, USA: John Wiley & Sons, Inc.
40. Ilk, N., Egelseer, E.M., Ferner-Ortner, J., Ku¨ pcu¨, S., Pum, D., Schuster, B., Sleytr, U.B. (2008) Surfaces functionalized with self-assembling S-layer fusion proteins for nanobiotechnological applications. *Colloids and Surfaces A: Physicochemical and Engineering Aspects* 321: 163–167.
41. Sara, M., Egelseer, E.M., Huber, C., Ilk, N., Pleschberger, M., Pum, D., Sleytr, U.B. S-layer proteins: potential application in nano(bio)technology. In: Rehm B, editor. *Microbial Bionanotechnology: Biological Self-Assembly Systems and Biopolymer-Based Nanostructures*, pp. 307–338. New Zealand: Bernd Rehm Massey University, PN; 2006..
42. Demir, M., Stowell, M.H. (2002) Viruses as self-assembled nanocontainers for encapsulation of functional cargoes. *Nanotechnology* 13: 541.
43. Ma, Y., Nolte, R.J., Cornelissen, J.J. (2012) Virus-based nanocarriers for drug delivery. *Advanced Drug Delivery Reviews* 64: 811.
44. Hung, C.W. (2008) *RNA Packaging and Gene Delivery using Tobacco Mosaic Virus Pseudo Virions*, Ph. D. Thesis, University of Maryland, US.
45. Ohtake, N., Niikura, K., Suzuki, T., Nagakawa, K., Mikuni, S., Matsuo, Y., Kinjo, M., Sawa, H., Ijiro, K. (2010) Gold nanoparticle arrangement on viral particles through carbohydrate recognition: A non-cross-linking approach to optical virus detection. *Chembiochemistry* 11: 959.
46. Minten, I.j., Claessen, V.I., Blank, K., Rowan, A.E., Nolte, R.J., Cornelissen, J.J. Catalytic capsids: the art of confinement. *Chemical Science* 2: 358.
47. Jung, B., Rao, A.L., Anvari, B. (2011) Optical nano-constructs composed of genome-depleted *brome mosaic virus* doped with a near infrared chromophore for potential biomedical applications. *ACS Nano* 5: 1243.
48. Miller, R.A., Presley, A.D., Francis, M.B. (2007) Self-assembling light-harvesting systems from synthetically modified tobacco mosaic virus coat proteins. *Journal of the American Chemical Society* 129: 3104.
49. Wang, Q., Chan, T.R., Hilgraf, R., Fokin, V.V., Sharpless, K.B., Finn, M.G. (2003) Bioconjugation by copper(I)-catalyzed azide-alkyne [3 + 2] cycloaddition. *Journal of the American Chemical Society* 125: 3192.

3 Synthesis of Nanomaterials via Physical, Chemical, and Biological Routes

Neeta Gupta, Arti Hadap, and Bhawana Jain

3.1 INTRODUCTION

The word "nano" dates back to the 1980s when scientists understood the necessity of examining things at the nanoscale. Even though the term nanotechnology was coined by Professor Norio Taniguchi in 1974, it had already been mentioned by Richard Feynman, who was a physicist, in his talk at an American Physical Society meeting which was held at the California Institute of Technology (Cal-Tech) in December, 1959 (*Nanotechnology*, n.d.). However, research on nanotechnology began after 1980 when the scanning tunnelling microscope was developed and it was possible to view atoms individually. Nanotechnology is a division of science that deals with producing materials which have a size less than 100 nm. Even though it is very hard to imagine something as tiny as the nanoscale, we can understand that a sheet of newspaper is about 100,000 nm thick. Nanotechnology can be understood as the ability to view and control discrete atoms and molecules. It can also be understood as the investigation and application of extremely small materials in various fields of science and technology.

Nanomaterials possess unique properties as a result of their exclusively large surface to volume ratio and quantum effects. Reducing the size of particles to nano size from micro size enables the material to acquire distinguished properties. Further tuning the particle morphology at this scale makes them suitable for versatile applications. On manipulating material at the nanoscale, the fundamental material properties such as magnetism, electricity, temperature, as well as chemical reactions can be changed entirely. Nanomaterials can exhibit nanoscale in 0D, 1D, 2D, and 3D dimensions. Nanomaterials exhibit unique optical, chemical, mechanical, electrical, magnetic, conductive, and various other properties. Some of the attractive properties of nanomaterials when compared to their larger counterparts are high material strength, light weight, advanced control over light spectrum, dimensional stability, and better chemical reactivity.

The methods used for synthesizing nanomaterials play a very crucial role in determining the overall properties of nanomaterials. The route used to synthesize materials largely determines the end properties of nanoparticles. Methods used for

DOI: 10.1201/9781003362258-3

synthesizing nanoparticles aid in tuning, tailoring, and imparting particular properties to a nanomaterial and make it suitable for particular applications. Hence a lot of time and energy is invested in developing methods for synthesizing nanomaterials.

The synthesis of nanomaterials has been classified into two categories for transforming raw materials to nanoscale materials: the top-down and bottom-up approaches (Ijaz et al., 2020). In the top-down approach, bulk material is cut down to produce materials having nanostructure, where initially the structure is large and it is externally processed to reduce its size. This is also referred to as a destructive method because here large/bulk molecules are disintegrated into tiny molecules which are later transformed into nanoparticles. This procedure involves extracting, tearing, and slicing large raw materials into nanoparticles. Some of the top-down methods are electro-explosion, stereolithography, sputtering, mechanical milling, electron beam etching, nanopatterning, subsurface reactive ion etching, and laser ablation.

In the bottom-up approach miniaturization of material components whose size varies up to the atomic level which are self-assembled to form the nanostructure. In this procedure the construction of the atomic structure is done by placement of atoms under particular circumstances in order to form the desired configuration (Satyanarayana, 2018). This method is also referred to as a constructive method as the nanoparticles are formed using simple substances. It is the opposite of the top-down approach. Some of the main methods used in this approach are chemical/vapor deposition and atomic layer deposition (ALD), others include the plasma-assisted technique, electrochemical deposition, sol–gel methods, ultrasonication, aerosol, etc. Figure 3.1 illustrates the two approaches used for synthesizing nanomaterials.

In this review we describe the various physical, chemical, and biological routes for the synthesis of nanoparticles. This review will be particularly useful for scientists and engineers who are looking to carry out research in developing novel routes for synthesizing nanomaterials.

3.2   SYNTHESIS OF NANOMATERIALS USING PHYSICAL ROUTES

3.2.1   CHEMICAL AND PHYSICAL VAPOR DEPOSITION

Physical vapor deposition (PVD) and chemical vapor deposition (CVD) are two processes used to deposit a very thin layer of material onto a substrate, known as a thin film (Selvakumar et al., 2012).

Because vapor deposition techniques produce products with superior hardness, wear resistance, smoothness, and oxidation resistance, they are the preferred processes for thin films. Thin films formed by vapor deposition are typically capable of functioning in unusual, high-stress environments.

PVD is used to produce articles that require films for mechanical, optical, chemical, or electronic functions such as semiconductors, thin-film solar panels, glass coatings, etc. In the meantime, CVD is used to create high-quality, high-performance solid materials. This method is widely used in the semiconductor industry to create thin films (Sadri, 2021). Due to low air pressure, nearly all PVD technologies have a poor coating performance on both the back and sides of the tool. To avoid shadow formation, the PVD reactor's loading density must be reduced, and loading and fixing

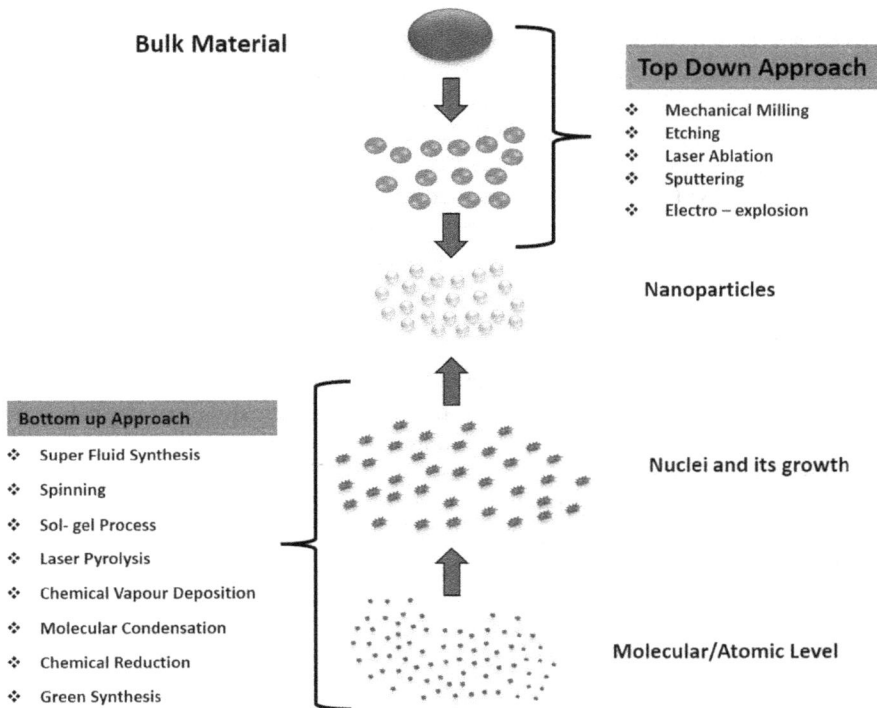

Bulk Material

Top Down Approach
- ❖ Mechanical Milling
- ❖ Etching
- ❖ Laser Ablation
- ❖ Sputtering
- ❖ Electro – explosion

Nanoparticles

Bottom up Approach
- ❖ Super Fluid Synthesis
- ❖ Spinning
- ❖ Sol- gel Process
- ❖ Laser Pyrolysis
- ❖ Chemical Vapour Deposition
- ❖ Molecular Condensation
- ❖ Chemical Reduction
- ❖ Green Synthesis

Nuclei and its growth

Molecular/Atomic Level

FIGURE 3.1 Description of top-down and bottom-up approaches for synthesizing nanomaterials.

are complicated. Meanwhile, the CVD method's process temperature is extremely high, often exceeding the tempering temperature of high-speed steel. As a result, in order to restore the hardness, the tools must be vacuum heat treated after coating.

PVD is a greener technology since it produces little pollution during the process because it involves a physical method. Meanwhile, the CVD reactive gas and reaction tail gas may be corrosive, flammable, and toxic, and the reaction tail gas may contain powdery and fragmented substances. Furthermore, the deposition temperatures differ between PVD and CVD, with PVD depositing at a relatively low temperature (around 250°C–450°C) and CVD depositing at relatively high temperatures ranging from 450°C to 1050°C. Further details on CVD and PVD has been summarized in Sections 3.2.1.1 and 3.2.1.2.

3.2.1.1   Chemical Vapor Deposition

CVD is a method of solid deposition and thin-film formation from vapor-phase materials. Although this method is somewhat similar to PVD, there are some differences between PVD and CVD. There are also different types of CVD such as laser CVD, photochemical CVD, low-pressure CVD, and organometallic CVD. In CVD, the coating material is on a substrate material, and it must be transported at a specific temperature in the form of vapor into a reaction chamber. There, the gas either decomposes and deposits on the substrate or reacts with it. A gas-delivery system, reacting chamber, substrate loading mechanism, and an energy source are thus required in a CVD apparatus. In order to ensure that only the reacting gas is present, the reaction also takes place in a vacuum. There must be a way to regulate the temperature and pressure inside the apparatus because, importantly, the substrate temperature is essential for determining the deposition. Finally, the equipment must have a way to dispose of excess waste gas. A volatile coating material which is stable to convert it to gas phase needs to be chosen as a carrier device to coat on the substrate (Gleason et al., 2010). Hydrides such as SiH_4, GeH_4, NH_3, halides, metal carbonyls, metal alkyls, and metal alkoxides are some of the precursors.

Volatile by-products of the entire CVD process must be safely removed by gas flow through the reaction chamber. Forced convection is used to move the source material into the reaction chamber, which also has a substrate, after it has been created. Diffusion causes reactants to be deposited into the substrate. The precursor eventually degrades, is removed from the substrate by diffusion, and leaves the desired layer of source material on the substrate after the mixture has adhered to it. Heat, plasma, or other methods can be used to facilitate or speed up the decomposition process (Tavares et al., 2008). Coatings, semiconductors, composites, nanomachines, optical fibers, and catalysts can all be made using the CVD technique.

The greatest advantages of using CVD are its ability to uniformly coat irregular surfaces such as threads and grooves. This process is very versatile since it is used in a wide variety of elements and compounds. CVD also produces thin films of very high purity and density. CVD is a relatively economical deposition process because many parts can be coated simultaneously.

3.2.1.2 Physical Vapor Deposition

PVD is primarily a vapor deposition coating technology in which the film of coating material is usually deposited atom by atom on a substrate by condensation from the vapor phase to the solid phase. However, it needs to perform the whole process under vacuum. First, a solid precursor material is bombarded with an electron beam, thereby ejecting atoms of the material. Second, these atoms enter the reaction chamber where the coating substrate is located. There, atoms can react with other gases during transport to form coating materials, or the atoms themselves can become coating materials. Finally, they stick to the substrate to form a thin layer. PVD coatings help reduce friction, make materials more resistant to oxidation, and increase hardness.

PVD is used as a deposition method to produce a very hard and corrosion-resistant coating. The thin film made of PVD has excellent high-temperature resistance and abrasion resistance. PVD uses physical processes such as sputtering and evaporation to produce vapors in the form of atoms, molecules, or ions from a coating material delivered by a target. They are then transported to the substrate surface and deposited to form a layer. In PVD processes, the substrate temperature is significantly below the melting temperature of the target material, allowing coating of temperature-sensitive materials. Examples of commonly used PVD processes include evaporation, ion plating, pulsed laser deposition, and sputter deposition. Compared to evaporation, sputtering is suitable for target materials that are difficult to deposit by evaporation, such as ceramics and refractory metals (Thian et al., 2006). In addition, sputtered coatings typically have better adhesion strength to substrates than evaporation deposited coatings (Zhong et al., 2019).

PVD has several advantages including: (i) the coating formed by PVD may have better properties than the substrate material; (ii) all types of inorganic materials and certain types of organic materials may be used; and (iii) the process is more environmentally friendly compared to many other processes such as electroplating. However, PVD also has several disadvantages including: (i) problems with coating complex shapes; (ii) high processing costs and low output; and (iii) process complexity (Makhlouf, 2011).

3.2.2 SONOCHEMICAL REDUCTION

In chemistry, the study of sonochemistry is concerned with understanding the effect of ultrasound on initiating or promoting chemical activity in a solution by forming acoustic cavitation in liquids (Einhorn et al., 1989). Therefore, the chemical effects of ultrasound are not caused by the direct interaction of ultrasound with molecules in solution.

There are three classes of sonochemical reactions, which are homogeneous sonochemistry of liquids, heterogeneous sonochemistry of liquid–liquid or solid–liquid systems, and sonocatalysis (increase or catalysis of chemical reaction rates by ultrasound) that overlaps with the previous ones (Pestman et al., 1994). Sonoluminescence is the result of the same cavitation phenomenon responsible for homogeneous sonochemistry (Crum, 1994). Chemical modification of reactions by ultrasound has beneficial applications in mixed-phase synthesis, materials chemistry, and biomedical applications. Since cavitation can only occur in liquids, no chemical reaction appears in ultrasonic irradiation of solid or solid–gas systems.

For example, in chemical kinetics, ultrasound has been observed to significantly increase chemical reactivity by a factor of one million in many systems, effectively activating heterogeneous catalysts (Suslick and Casadonte, 1987). In addition, for reactions at the liquid–solid interface, ultrasound breaks up solid pieces, revealing an active and clean surface of microjet holes from near-surface cavitation and solid fragmentation from near-surface cavitation collapse. This gives the solid reactants a larger surface area than the active surface for the reaction to take place, increasing the rate of the observed reaction (Zeiger and Suslick, 2011; Hinman and Suslick, 2017).

The use of ultrasonic irradiation to reduce the antioxidant activity of olive oil mill wastewaters (OMWs) derived from two-phase and three-phase decanters has been investigated. Sonication of diluted OMW samples was performed at ultrasonic frequencies of 24 and 80 kHz, applied power varied between 75 and 150 W, and liquid bulk temperature varied between 25 and 60°C. Antioxidant activity was found to increase with decreasing temperature, increasing in power and frequency. Addition of NaCl to the sample also appears to increase the reduction. The antioxidant activity of OMW samples was evaluated using a newly developed Co(II)/ethylenediaminetetraacetic acid (EDTA) derivatized luminol chemiluminescence spectroscopy protocol, and total phenol loading was determined according to the Folin-Ciocalteu method (Atanassova et al., 2005).

3.2.3 Laser Ablation

Laser ablation synthesis produces nanoparticles by striking the target material with an intense laser pulse. The word "ablation" indicates the method of eliminating surface atoms by breaking chemical bonds, which incorporates both a single-photon process and multiphoton stimulation (thermal evaporation) (Baig et al., 2021). During pulse laser ablation (PLA), the source material or precursor vaporizes due to the laser's high energy, causing nanoparticle production. Early in the 1990s, Fojtik and Henglein's work was the first report on the PLA on colloidal liquid synthesis of nanomaterials (Fojtik et al., 1994). Over the years, this technique has emerged as a significant study area for both metal and non-metal nanoparticle applications. This method can be used to create a wide variety of nanomaterials, including metal nanoparticles, ceramics, carbon nanomaterials, and oxide composites (Baig et al., 2021).

An example of the PLA experimental setup is shown in Figure 3.2. At the base of the reaction chamber, the samples were positioned on a spinning sample holder. A focusing lens focused the laser beam at a 45° grazing angle on the target surface. The flow of N_2 gas was established along the chamber to be able to guide and regulate the ablated particles. The ablated particles either went straight into the particle size analyzer at the gas stream's outflow or were collected for chemical analysis on a filter (Gera et al., 2020).

The desired nanoparticle properties, such as size distribution, can be achieved by adjusting the parameters of the laser pulse (e.g., wavelength and fluence) (Baig et al., 2021). According to Forsythe and co-workers (Forsythe et al., 2021), the average size

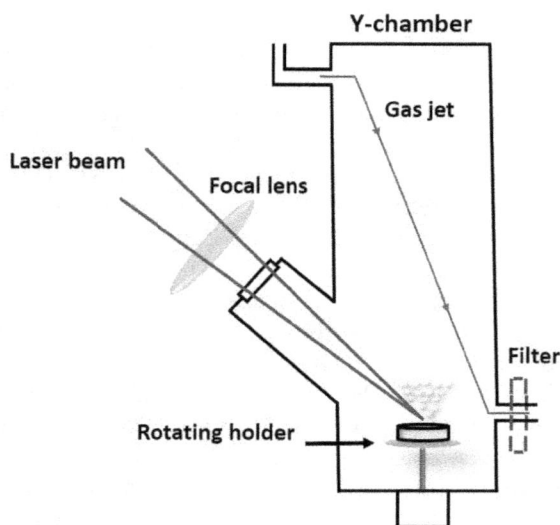

FIGURE 3.2 Pulsed laser ablation experimental setup.

and distribution of the nanoparticles produced by varying the laser parameters can be controlled in practice.

PLA has many advantages over traditional approaches including simplicity, safety, less porosity, the absence of a surfactant requirement, quick repetition rates, the capacity to produce materials with complex stoichiometry, and narrower particle size distribution. In addition, this method is cheap, simple, and requires a small number of chemical species, making it effective for producing high-purity nanoparticles because only metal targets and water are required for preparation (Meidanchi & Jafari, 2019; Baig et al., 2021). Due to these benefits, laser ablation techniques are the most significant and widely used methods for creating nanoparticles and depositing films (Sportelli et al., 2018). Compared to other conventional methods, laser ablation also allows the selective control of phase composition, size, and shape of the nanoparticles, as reported by Zhang and co-workers (Zhang et al., 2021).

Although regarded as an effective technology, PLA has several downsides (Sportelli et al., 2018). Even though PLA allows the uniform distribution of nanomaterials and precise chemical control, it is costly to use. Furthermore, managing a laser state with such independent precision and energy requirements is challenging. After all, delivering good ablation efficiency requires a significant quantity of energy. Ablation efficiency is reduced with prolonged ablation times due to the large number of nanoparticles positioned along the laser beam (Baig et al., 2021; Cui et al., 2021). This problem can be avoided by adopting a flow-through device to remove the nanoparticles from the optical path.

A wide range of nanomaterials can be produced through these PLA techniques. For example, Meidanchi and Jafri synthesized Ta_2O_5 nanoparticles using the beam of a Q-switched Nd:YAG laser to irradiate the tantalum metal target in water (Meidanchi & Jafari, 2019). The findings demonstrate that the high purity of Ta_2O_5 nanoparticles was produced in orthorhombic crystalline structure and spherical shape with average diameters of around 5–20 nm. Furthermore, Yudasari and co-workers successfully synthesized bi-phase Zn/ZnO nanoparticles by PLA in pure water and *P. pinnata* leaf extract solution. The Zn/ZnO nanoparticles produced have a spherical shape with a smaller size in comparison to pure ZnO (Yudasari et al., 2021). In addition, magnetite nanoparticles also can be produced by PLA using Nd:YAG (neodymium-doped yttrium aluminum garnet), Ti:sapphire (titanium-doped sapphire), and copper vapor lasers.

3.2.4 PLASMA

One of the most common methods for producing nanomaterials is plasma synthesis. Plasma is regarded to be the fourth state of matter in addition to solid, liquid, and gas. Physical plasma is a part of ionized gas that consists of electrons, ions, excited species, electric fields, reactive species, UV photons, etc. (Chang, 2013; Adamovich et al., 2017). The topic of plasma synthesis of nanomaterials and nanostructured materials has considerably increased in relevance in fields including biology, medicine, energy conversion, and electronics. Furthermore, the production of nanomaterials has sparked tremendous interest in plasma nanofabrication (Kaushik et al., 2019a). Plasma production in the fabrication of nanomaterials can occur naturally or be performed in laboratories. Thermal plasma technologies offer a versatile and scalable platform for creating nanomaterials, including oxides, carbides, nitrides, and composites.

According to D'Angola et al. (2022), the emergence of plasma technology in producing nanomaterials is classified into two categories, which are thermal and non-thermal plasma synthesis. The relative temperatures of electrons, ions, and ionic compounds determine whether the plasma is thermal or non-thermal. The term thermal (or "hot") plasma refers to a state of matter when the temperatures of the electrons and heavy particles have reached thermal equilibrium or are kept constant. When the electron temperature is significantly higher than that of the ions and neutral particles, a non-thermal (or "cold") plasma is created (e.g., usually at room temperature). Therefore, thermal, and non-thermal plasmas could produce ionizing gases, even though the temperature of the plasma species is not the same, by lowering or raising the temperature, allowing nanomaterials to be synthesized (D'Angola et al., 2022; Nava-Avendaño et al., 2021).

Figure 3.3 shows the experimental setup of plasma synthesis. Inductively coupled radiofrequency (RF) discharge is used by the plasma reactor to maintain the relatively uniform and dense plasma in nitrogen. Heating coils are wrapped around an evacuated chamber to generate plasma by heating the metal above its evaporation point (Chang, 2013). When nitrogen, N_2, gas is introduced into the system, a high-temperature plasma near the coils is produced. Metal vapor forms on nitrogen gas and diffuses to a cold collector rod, where nanoparticles are collected, and oxygen gas is used to passivate them. Optical emission spectroscopy (OES) was used to further characterize plasma (Vesel et al., 2021a).

This method is more environmentally friendly and cost-effective than chemical processes to achieve the desired structural and electronic properties. The first is non-thermal plasma nanomaterial synthesis, which has been demonstrated to be a simple,

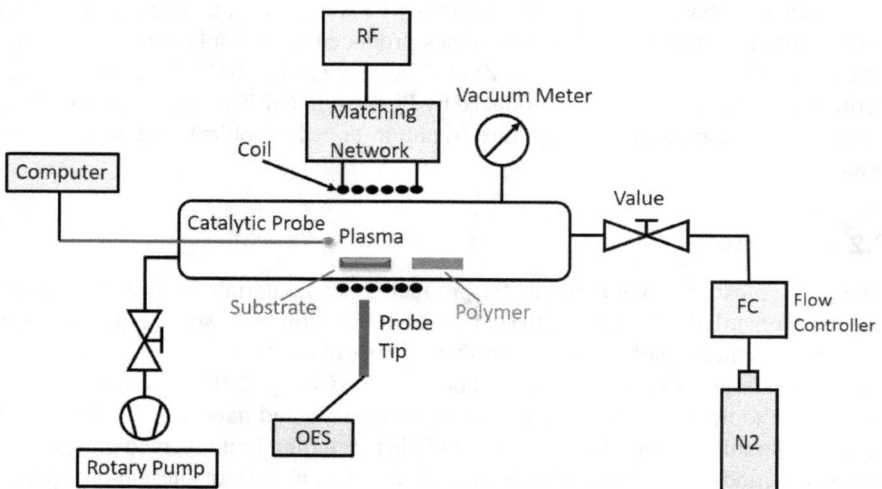

FIGURE 3.3 Plasma experimental setup.

effective, low-cost, and clean method (Kaushik et al., 2019). It has the potential to reduce the use of toxic chemical compounds in the manufacturing process. As a result, the toxicity of nanomaterials synthesized using plasma is reduced. In terms of the number of nanomaterials produced, plasma pyrolysis is significantly more expensive (Luo et al., 2019). The plasma fabrication method provides the least amount of flexibility in terms of nanoparticle design, precision, and customization.

Nevertheless, the main drawback of this method is that the size of the nanoparticles produced by this method is difficult to control, and many particles have a strong tendency to aggregate. Because of nanoparticle aggregation within the bulk powder, surface functionalization can be difficult to impossible. This limits the ability to produce high-quality nanoparticle dispersions. Other aspects of nanoparticle design are severely limited, such as composition and morphology, as well as the ability to create complex structures such as alloys and doped structures.

Recently, Nava-Avendaño and co-workers demonstrated the synthesis of Li_2S using thermal plasma technology. The preparation of nanosized Li_2S from sulfur and lithium hydroxide, as well as the carbothermal reduction of lithium sulfate, was accomplished using an inductively coupled plasma reactor (Nava-Avendaño et al. 2021). Furthermore, Vesel and co-workers used direct plasma to create N-doped carbon nanomesh. Nitrogen radicals (plasma) reacted with the polymer as a carbon source, causing the nitrogen–carbon fragments to desorb on the Ti substrate. This effect allowed a layer of N-doped carbon nanomesh to form on the substrate (Vesel et al., 2021b). In addition, Primc and co-workers demonstrated the synthesis of ZnO nanoparticles using nonequilibrium gaseous plasma. The precursors were introduced directly into the gaseous plasma and plasma reactors operating at low pressures to avoid agglomeration of the synthesized ZnO nanoparticles (Primc et al., 2021).

3.3   NANOMATERIAL SYNTHESIS USING CHEMICAL ROUTES

3.3.1   POLYOL METHOD

The polyol method is a versatile liquid-phase synthesis route that uses high boiling and multivalent alcohols (polyols) to synthesize nanomaterials. The formation of particle using the polyol method involves the reduction of the solid precursor by polyol in the presence of capping agents. Various nanomaterials have been synthesized using the polyol method, including metal nanoparticles, nanostructures, nanoplates, nanowires, and metal nanorods.

The polyol method was first discovered in 1989 by Fievet et al. to synthesize well-defined metal nanoparticles from their respective oxides, hydroxides, or salts in polyols. It was first started with cobalt, nickel, and copper spherical particles and then extended to noble metal particles (rhenium, ruthenium, rhodium, palladium, silver, platinum, and gold) and polymetallic particles (CoNi, FeNi FeCo, and FeCoNi).

The reaction occurs through the total dissolution of the solid precursor in the polyol solvent, the precipitation of a solid from a solution, homogeneous nucleation, and particle growth of nuclei. The polyols act as solvents of the solid precursor and a reducing and stabilizing agent to control particle growth and prevent aggregation.

Several critical parameters, including polyol solvents, capping agents, reducing agents, metal precursor concentrations, and atomic species, can affect particle formation sizes, shapes, compositions, and crystallinity. Several polyols have been used, including ethylene glycol, diethylene glycol, triethylene glycol, tetraethylene glycol, 1,3-propane diol, 1,4-butane diol, 1,5-pentane diol, and glycerol. The most commonly used polyols are ethylene glycol, diethylene glycol, 1,4-butane diol, and glycerol. The selection of polyols depends on the optimum reflux temperature and the reduction potential of polyols. Therefore, the proper selection and precise control of all these parameters are critical for synthesizing nanomaterials with well-defined/controlled sizes and shapes.

The polyol method provides many advantages. These advantages include: (i) polyols have high boiling points, allowing the synthesis of well-crystallized materials at relatively high temperature without applying high pressure; (ii) good capability of polyols to dissolute common metal salts, allowing the use of simple and cheap metal precursors as stating compounds; (iii) high viscosities; (iv) reductive ability at elevated temperatures for the direct synthesis of metal; (v) complexing agents; and (vi) chelating ability of polyol solvent allows flexibility in controlling the size, texture, and shape of the final particles.

3.3.2 MICROEMULSION

The microemulsion, a subset of emulsions, was first coined by Schulman et al. in 1959. It is a dispersion system of two immiscible fluids mixed and stabilized by an interfacial film of surfactant(s). This isotropic system is a versatile and reproducible preparation technique. The main advantage of the microemulsion is the distinct ability to control the properties of particles, which include size, geometry, morphology, homogeneity, and surface area. It is a suitable route for synthesizing size-controlled nanomaterials. Many nanomaterials have been synthesized using the microemulsion method, which includes metal nanoparticles, metal oxide nanoparticles, magnetic nanoparticles, and nanocomposites.

Its dispersed domains are generally in the nanometer size range (1–100 nanometers in diameter). Therefore, microemulsion appears as a transparent or translucent solution. Their visual appearance does not change with time. Hence, it is considered a single-phase (monophasic) thermodynamically stable system.

Microemulsion consists of at least three components, namely a dispersed phase (polar phase), a continuous phase (non-polar phase), and surfactant(s) with or without co-surfactant. The co-surfactant is usually short-chain alcohol. The surfactant molecule comprises a hydrophilic headgroup and a long hydrophobic carbon tail. At the microscopic level, surfactant molecules form an interfacial film separating the polar and non-polar domains. This interfacial layer forms different microstructures.

The main types of microemulsion are (i) water-in-oil (W/O) microemulsion when the water-soluble molecules are dispersed within a continuous oil phase; (ii) oil-in-water (O/W) microemulsion when the oil is dispersed in a continuous water phase; and (iii) bicontinuous microemulsion. Among these microemulsions, W/O microemulsion has been most extensively studied.

There are different stages of particle formation inside droplets: chemical reaction, nucleation, and particle growth. In microemulsion, the most common method is the two-microemulsion method, which is based on a fusion–fission event between the droplets. Two different reactants are introduced in two identical microemulsions. These two microemulsions are then mixed up by constant mixing. After mixing both emulsions, the droplets continuously collide, coalesce, and break apart, resulting in a continuous exchange of solution contents. Then, the reaction can occur inside the droplets, acting as nanoreactors, which can be used to synthesize nanomaterials with low polydispersity. The chemical reaction is initiated when there is a fusion–fission event between the droplets. After the initiation of chemical reaction at the droplets, this will lead to particle formation that eventually leads to nuclei formation and subsequently to the growth of particles. In other methods, one of the reactants is introduced in solution into a microemulsion containing another reactant or is added directly to the microemulsion.

The different parameters involved in particle formation in microemulsion have greatly influenced their size and polydispersity. These parameters include reactant exchange rate constant, film flexibility, and reactant concentration. The careful selection of specific parameters and the appropriate control of all these parameters can control particle size and provide homogeneous distribution of particle sizes.

3.3.3 Thermal Decomposition

Thermal decomposition is an endothermic process in which the solid bulk materials are decomposed by breaking the chemical bond into its constituents caused by heat (Baig et al., 2021). This process is among the simplest, most cost-effective, and most practical methods for producing nanoparticles. Additionally, it addresses the foremost challenge in nanotechnology research by providing a solution to obtaining controlled nanometric size and shape (Odularu, 2018).

Figure 3.4 shows the experimental setup of thermal decomposition synthesis. The water-cooled reflux aperture electronically coupled heating mantle (heat), a peristaltic pump (sample out), X-ray transparent Kapton sample chamber, and autosampler were all used in the experiments. An electrically connected heating mantle warmed the water-cooled reflux aperture. The sample was continuously pushed from the reaction mixture by a peristaltic pump through the X-ray transparent Kapton sample chamber and sampled by an autosampler (Lassenberger et al., 2017).

This process typically requires several chemical substances, including the precursor, a solvent, and stabilizing surfactants such oleic acid and oleylamine. Nanoparticles can be synthesized from a precursor of various metallic compounds, including coordination complexes, organometallic compounds, and alkoxides. Furthermore, high boiling point solvents with working temperatures between 100 and 350°C are used in thermal decomposition procedures, including octadecene, oleic acid, kerosene, octyl ether, oleylamine, and polyethylene glycol (Palacios-Hernández et al., 2012).

The concentrations of metal precursors, surfactants, reducing agents, and temperature all influence the rate of the reaction process (Baig et al., 2021; Maity et al., 2009). Controlling the time, temperature, reagent concentration, and surfactant type can

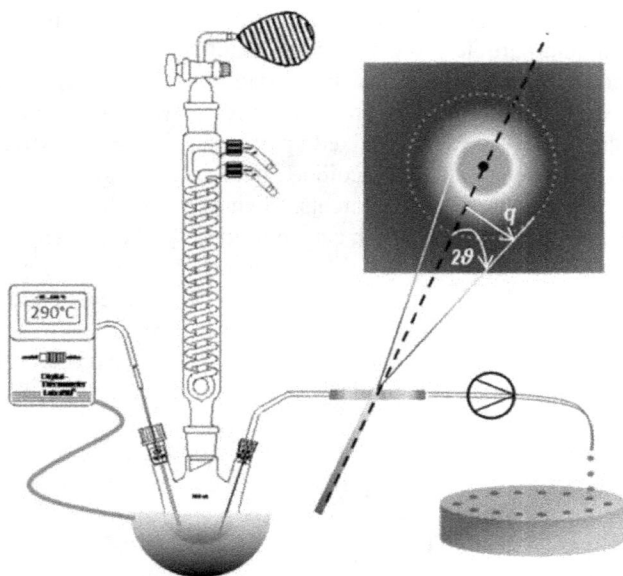

FIGURE 3.4 Thermal decomposition experimental setup.

result in uniform growth and size distribution of nanoparticles. In addition, thermal decomposition takes place at a low reaction temperature which yields particles with extremely narrow size distributions and no byproducts (Saima et al., 2019).

In large-scale synthesis, solventless procedures are more attractive because they are simpler to use, generate greater yields, create fewer byproducts, and, with careful temperature and decomposition condition control, may achieve smaller sizes without nanoparticle aggregation. Therefore, it demands the use of inexpensive precursors, a straightforward procedure, and no further purification steps in the industry (Palacios-Hernández et al., 2012). Recently, Deshmukh and co-workers demonstrated a simple and extensive method for producing TiO_2 nanoparticles for large-scale synthesis by thermal decomposition of titanium oxysulfate with urea (Deshmukh et al., 2021). They highlight the function of urea in controlling the TiO_2 nanoparticles' size, shape, and aggregation during thermal decomposition.

3.3.4 ELECTROCHEMICAL SYNTHESIS

Electrochemical synthesis is the synthesis of nanomaterials in an electrochemical cell. Fundamentally, electrochemical synthesis can be described as the transfer of an electric current between two or more electrodes, referred to as the anode and cathode, which are placed in an electrolyte. In this method, the anode will be oxidized to metal ion species in the electrolyte, and the metal ion is then reduced to metal by the cathode with the help of stabilizers. The synthesis basically takes place at the electrode–electrolyte interface. The product formed is coated on the electrode as

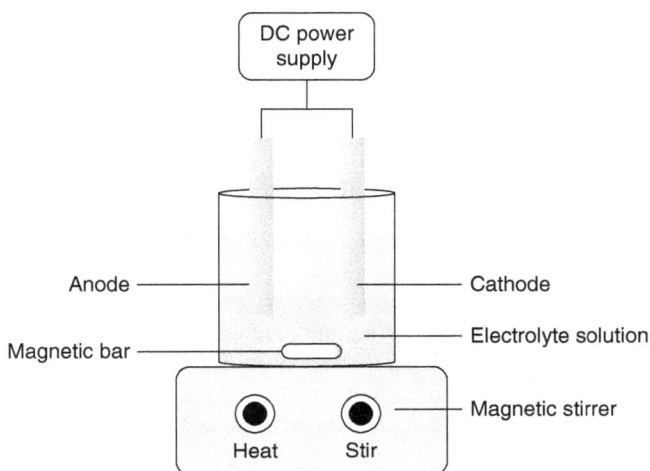

FIGURE 3.5   The experimental setup of electrochemical synthesis.

Source: Ramimoghadam et al., 2014.

a coating or thin films with various morphologies (Ramimoghadam et al., 2014). Figure 3.5 shows the basic experimental setup for electrochemical synthesis.

The electrochemical synthesis of nanomaterials depends on the electrode material, the electrochemical window of the electrolyte, the type of precursor, additives, fundamental electrochemical parameters, and temperature. By changing the experimental conditions, electrochemical technologies enable one-step synthesis of nanomaterials (Lebedeva et al., 2021; Singaravelan & Bangaru, 2015). In addition, by altering the composition of the electrolyte, it is possible to regulate the film structure (Ramimoghadam et al., 2014) and when combining the electrochemical with the template, it allows the creation of 3D networks, such as mesoporous silica film (Tonelli et al., 2019).

This method has many advantages due to being simple to carry out, with a rapid synthesis time, and the equipment being affordable and readily available. Typically, these electrochemical reactions result in products that cannot be produced through chemical synthesis. The oxidizing or/and reducing power may be continuously varied and suitably adjusted in the electrochemical synthesis method since it is based on an oxidation/reduction reaction; this capability cannot be achieved by other chemical synthesis techniques (Ramimoghadam et al., 2014; Li et al., 2013).

However, there are several drawbacks to this approach that must be considered. Since reactions occur at ambient temperature, the electrochemical synthesis frequently results in disordered products that make precise structural characterization more challenging. During X-ray characterization, synthesized materials typically exhibit amorphous impurities.

The synthesis of nanoparticles without the use of surfactants has recently been described using a novel electrochemical approach. This procedure used aqueous nanodroplets (AnDs, in continuous-phase organic solvent) containing precursor ions to the desired nanoparticle. These AnDs undergo electrolysis when they come into contact with an electrode, which promotes the synthesis of nanoparticles (Park & Ahn, 2020)

3.4 NANOMATERIAL SYNTHESIS USING THE BIOLOGICAL ROUTE

The biological route for the synthesis of nanoparticles employs using microorganisms such as fungi and bacteria and plant extracts. The biological route has caught the attention of researchers making nanomaterials, and is known as green synthesis. Green synthesis is one of the bottom-up approaches to replacing chemical reducing agents with biological entities. This method provides several advantages, including wide availability, low cost, environmentally friendly, non-toxicity, biocompatibility, time saving, and good stability, as shown in Figure 3.6. These nanomaterials are referred to as biogenic nanomaterials.

Plant extracts (flower, leaf, root, seed, bark, petals, fruits, and peels) and microorganisms (fungi, bacteria, algae, yeast, and some viruses) are used in the green synthesis of bionanomaterials. In most biological processes, however, the mechanism is only vaguely understood. The size and shape of the resulting nanomaterials are usually controlled by the properties of the biological entity, and organic reducing agents also play a major role (Aswathy Aromal & Philip, 2012). Other factors influencing nanoparticle structural properties include pH, target salt concentration, temperature (room temperature or required external heat sources), agitation conditions, time variation, and biological reducing agent (Hartman et al., 2019). The production of bionanomaterials by the green synthesis method occurs either intracellularly or extracellularly depending on the location of the released biomolecules.

3.4.1 Plant Extract-Based Nanoparticle Synthesis

Plants are considered to be more suitable than microorganisms for the green synthesis of nanoparticles. This is due to the fact that plants have more advantages, such as

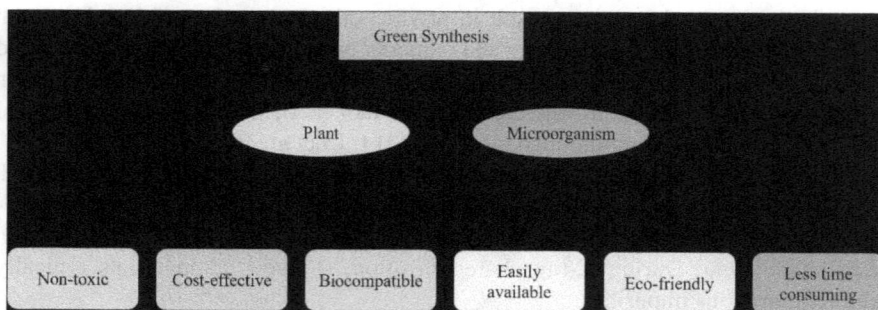

FIGURE 3.6 Advantages of the green synthesis method.

being non-pathogenic, having a one-step protocol synthesis, having shorter reaction times, and having various pathways that have been thoroughly researched (Baker et al., 2013). Plant research has made remarkable progress in recent years, with the current surge in plant research focusing on the synthesis of nanoparticles with controlled size and shape. Plants are used to produce a wide range of metal and metal-oxide nanoparticles. When compared to their bulk counterparts, these nanoparticles have distinct optical, thermal, magnetic, physical, and electrical properties, and they have numerous applications in a wide range of human interests.

Enzymes, vitamins, amino acids, polysaccharides, proteins, polyphenols, and organic acids (such as citrate) found in plant extracts can reduce metal ions to form nanomaterials. Biomolecules, in addition to reducing ions, play an important role in capping nanoparticles and stabilizing the formed nanoparticles to the desired size and shape (Baker et al., 2013). Plant parts such as leaves, roots, flowers, stems, seeds, and fruit have been used to create bionanomaterials of various sizes and shapes (Table 3.1).

3.4.2 Microbial Synthesis of Nanoparticles

Microorganisms are extremely important due to their increased genetic diversity. Various bacteria, fungi, yeast, algae, and viruses have been discovered and are primarily used in the synthesis of silver, gold, platinum, titanium dioxide, zinc oxide, and other metals. In comparison to using chemical and physical methods over time, it is less expensive, less cumbersome, requires less energy, wastes less input, has more practical control of constituent ingredients, is non-toxic, and has excellent yield (Ramrakhiani et al., 2016).

Depending on the type of microorganism used, nanoparticle formation by microorganisms can be extracellular or intracellular. Extracellular synthesis of nanoparticles, on the other hand, is more advantageous because it does not lyse the host cells and allows for easy product recovery via simple down-streaming (Kuppusamy et al., 2016). Table 3.2 lists some microorganisms that have been used in the manufacture of nanomaterials.

3.5 ISSUES AND CHALLENGES IN SCALING UP PRODUCTION OF NANOMATERIALS

In order to replace the prevailing materials with nanoparticles that display better performance with better capabilities, mass production of nanomaterials is required. Nanoscience provides a one-of-a-kind opportunity to realize the physiochemical facts at the utmost elementary state. Theoretically, these findings could prove valuable for the improvement and optimization of a structure under consideration.

Apart from these rudimentary achievements, nanotechnology is facing considerable limitations. One of the major challenges is converting the scientific data and research published in standard journals into real-time industrial practices and applications. This issue is complicated. Firstly, there are significant changes in material characteristics on scaling up just as when they are scaled down to the nanoscale. Particularly, the

.

.

.

.

.

TABLE 3.1
Different Plants Produce Different Types of Nanomaterials for Use in Various Applications

Plant	Nanomaterial	Morphology	Application	References
Gloriosa superba L.	Copper oxide	5–10 nm, spherical	Antibacterial agent	Naika et al. (2015)
Aloe vera	Indium oxide	5–50 nm, cubic	Solar cells, gas sensors	Maensiri et al. (2008)
Sida acuta	Chitosan-cerium oxide	23.12–89.91 nm, spherical	Antibacterial agent	Senthilkumar et al. (2017)
Mulberry	Silver	15–25 nm, spherical	Antimicrobial agent	Singh et al. (2017)
Lagenaria Siceraria	Iron oxide	30–100 nm, cubic	Antimicrobial agent	Kanagasubbulakshmi & Kadirvelu (2017)
Gardenia resinifera	Iron oxide	Around 5 nm, spherical	Hyperthermia	Karade et al. (2019)
Magnolia kobus	Copper	37–110 nm, spherical	Antibacterial agent	Lee et al. (2013)
Hibiscus rosa-sinensis	Iron oxide	65 nm	Fortifying wheat biscuits	Razack et al. (2020)
Plantago major	Iron oxide	4.6–30.6 nm, spherical	Catalyst	Lohrasbi et al. (2019)
Arbutus Unedo	Silver	30 nm, spherical	Biomedical	Kouvaris et al. (2012)
Red ginseng	Silver and gold	10–30 nm, spherical	Biomedical	Singh et al. (2016)

TABLE 3.2
Types of Nanomaterials Produced by Various Microorganisms in Different Applications

Microorganism	Nanomaterial	Morphology	Application	References
Bacillus mycoides	Titanium dioxide	40–60 nm, spherical	Quantum dot sensitized solar cells (QDSSCs)	Órdenes-Aenishanslins et al. (2014)
Humicola sp.	Cerium oxide	12–20 nm, spherical	Biomedical	Khan & Ahmad (2013)
Yeast cells	Gold–silver alloy	9–25 nm	Vanillin sensor	Zheng et al. (2010)
Bacillus subtilis	Selenium	50–400 nm, spherical	H_2O_2 biosensor	Wang et al. (2010)
Saccharomyces cerevisiae	Selenium sulfide	6.0–153 nm, spherical	Skin disease	Asghari-Paskiabi et al. (2019)
Portieria hornemannii (red algae)	Silver	60–70 nm, Spherical	Antibacterial activity against fish pathogens	Fatima et al. (2020)
Streptomyces sp.	Zinc oxide	20–50 nm, spherical	Anticancer	Balraj et al. (2017)
Colpomenia sinuosa and *Pterocladia capillacea*	Iron oxide	11.24–33.71 nm and 16.85–22.47 nm, spherical	Antimicrobial agent	Salem et al. (2019)
Exiguobacterium mexicanum	Silver	5–40 nm	Biomedicine	Padman et al. (2014)

control that can be exercised at the scale of nanometer inclines to diminish at scales of macro and meso-.

Also, industries are reluctant to invest considerably on the development of upcoming large-scale technologies for manufacturing of nanomaterials unless a huge amount of profit is guaranteed. This predicament is predominantly significant in applied sciences because of the gap between the lab-scale and industrial-scale research work.

Although there is a lot of understanding and data on the potential of nanoparticles to empower a broad series of innovations across industries, expertise in the efficient transition of nanomaterials for commercial use remains lacking. Nanomaterials need to have a process for commercialization which on careful execution will lead to successful development of systems and processes.

3.6   CONCLUSION

Nanotechnology has a great impact on technologies associated with electronics, optics, automobiles, construction, biomedical devices, medical implants, and various others. This is possible as nanomaterials exhibit various unique and attractive features that find application in engineering and technology. Most of these nanomaterials have already been employed for commercial applications which include sunscreens, drug-delivery systems, automobiles, tissue regeneration, etc. However, many applications are still under research and hold huge potential for commercialization which would be a boon to humanity.

Here, in this review chapter, we have described in detail the different methods used for the synthesis of nanoparticles. Nanoparticles can be synthesized primarily using two approaches: the top-down and bottom-up. We have also described in detail the various physical, chemical, and biological routes for synthesis. We have also mentioned the various challenges encountered in the scaling up of nanomaterials. Scaling up requires developing an efficient and cost-effective strategy for producing nanomaterials which will be suitable for various industrial applications. Hence, a standard procedure which has been carefully examined, developed, tried, and tested for transferring of nanomaterials from lab to industry should be adopted. This must ensure a cost-effective and perfect method for the production of nanomaterials that can be used for different applications.

REFERENCES

Adamovich, I., Baalrud, S.D., Bogaerts, A., Bruggeman, P.J., Cappelli, M., Colombo, V., Czarnetzki, U., Ebert, U., Eden, J.G., Favia, P., Graves, D.B., Hamaguchi, S., Hieftje, G., Hori, M., Kaganovich, I.D., Kortshagen, U., Kushner, M.J., Mason, N.J., Mazouffre, S., Vardelle, A. (2017) The 2017 Plasma Roadmap: Low temperature plasma science and technology. *Journal of Physics D: Applied Physics* 50(32): 323001.

Alf, M.E., Asatekin, A., Barr, M.C., Baxamusa, S.H., Chelawat, H., Ozaydin-Ince, G., Petruczok, C.D., Sreenivasan, R., Tenhaeff, W.E., Trujillo, N.J., Vaddiraju, S., Xu, J., Gleason, K.K. (2009) Chemical vapor deposition of conformal, functional, and responsive polymer films. *Advanced Materials* 22(18): 1993–2027.

Asghari-Paskiabi, F., Imani, M., Rafii-Tabar, H., Razzaghi-Abyaneh, M. (2019) Physicochemical properties, antifungal activity and cytotoxicity of selenium sulfide nanoparticles green synthesized by *Saccharomyces cerevisiae*. *Biochemical and Biophysical Research Communications* 516(4): 1078–1084.

Aswathy Aromal, S., Philip, D. (2012) Green synthesis of gold nanoparticles using Trigonella foenum-graecum and its size-dependent catalytic activity. *Spectrochimica Acta Part A: Molecular and Biomolecular Spectroscopy* 97: 1–5.

Atanassova, D., Kefalas, P., Petrakis, C., Mantzavinos, D., Kalogerakis, N., Psillakis, E. (2005) Sonochemical reduction of the antioxidant activity of olive mill wastewater. *Environment International* 31(2): 281–287.

Baig, N., Kammakakam, I., Falath, W. (2021) Nanomaterials: A review of synthesis methods, properties, recent progress, and challenges. *Materials Advances* 2(6): 1821–1871.

Baker, L.A., Ueberheide, B.M., Dewell, S., Chait, B.T., Zheng, D., Allis, C.D. (2013) The yeast Snt2 protein coordinates the transcriptional response to hydrogen peroxide-mediated oxidative stress. *Molecular and Cellular Biology* 33(19): 3735–3748.

Baker, S., Rakshith, D., Kavitha, K.S., Santosh, P., Kavitha, H.U., Rao, Y., Satish, S. (2013) Plants: Emerging as nanofactories towards facile route in synthesis of nanoparticles. *BioImpacts* 3(3): 111–117.

Balraj, B., Senthilkumar, N., Siva, C., Krithikadevi, R., Julie, A., Potheher, I.V., Arulmozhi, M. (2017) Synthesis and characterization of zinc oxide nanoparticles using marine *Streptomyces* sp. with its investigations on anticancer and antibacterial activity. *Research on Chemical Intermediates* 43(4): 2367–2376.

Crum, L.A. (1994) Sonoluminescence, sonochemistry, and sonophysics. *The Journal of the Acoustical Society of America* 95(1): 559–562.

Cui, S., McClements, D.J., Shi, J., Xu, X., Ning, F., Liu, C., Zhou, L., Sun, Q., Dai, L. (2023) Fabrication and characterization of low-fat Pickering emulsion gels stabilized by zein/phytic acid complex nanoparticles. *Food Chemistry* 402: 134179.

D'Angola, A., Colonna, G., Kustova, E. (2022) Editorial: Thermal and non-thermal plasmas at atmospheric pressure. *Frontiers in Physics* 10: 852905.

Einhorn, C., Einhorn, J., Luche, J.-L. (1989) Sonochemistry—The Use of Ultrasonic Waves in Synthetic Organic Chemistry. *Synthesis*,1989(11): 787–813.

Fatima, R., Priya, M., Indurthi, L., Radhakrishnan, V., Sudhakaran, R. (2020) Biosynthesis of silver nanoparticles using red algae *Portieria hornemannii* and its antibacterial activity against fish pathogens. *Microbial Pathogenesis* 138(October 2019): 103780.

Fojtik, A., Henglein, A. (1994) Luminescent colloidal silicon particles. *Chemical Physics Letters* 221(5–6): 363–367.

Forsythe, R.C., Cox, C.P., Wilsey, M.K., Müller, A.M. (2021) Pulsed laser in liquids made nanomaterials for catalysis. *Chemical Reviews* 121(13): 7568–7637.

Gera, T., Nagy, E., Smausz, T., Budai, J., Ajtai, T., Kun-Szabó, F., Homik, Z., Kopniczky, J., Bozóki, Z., Szabó-Révész, P., Ambrus, R., Hopp, B. (2020) Application of pulsed laser ablation (PLA) for the size reduction of non-steroidal anti-inflammatory drugs (NSAIDs). *Scientific Reports* 10(1): 15806.

Hartman, S., Liu, Z., van Veen, H., Vicente, J., Reinen, E., Martopawiro, S., Zhang, H., van Dongen, N., Bosman, F., Bassel, G.W., Visser, E.J.W., Bailey-Serres, J., Theodoulou, F.L., Hebelstrup, K.H., Gibbs, D.J., Holdsworth, M.J., Sasidharan, R., Voesenek, L.A.C.J. (2019) Ethylene-mediated nitric oxide depletion pre-adapts plants to hypoxia stress. *Nature Communications* 10(1): 1–9.

Hinman, J.J., Suslick, K.S. (2017) Nanostructured materials synthesis using ultrasound. In: J.C. Colmenares, G. Chatel (Eds.) *Sonochemistry: From Basic Principles to Innovative Applications* (pp. 59–94). Springer International Publishing.

Ijaz, I., Gilani, E., Nazir, A., Bukhari, A. (2020) Detail review on chemical, physical and green synthesis, classification, characterizations and applications of nanoparticles. *Green Chemistry Letters and Reviews* 13(3): 59–81.

Kanagasubbulakshmi, S., Kadirvelu, K. (2017). Green synthesis of Iron oxide nanoparticles using Lagenaria siceraria and evaluation of its antimicrobial activity. *Defence Life Science Journal* 2(4): 422.

Karade, V.C., Parit, S.B., Dawkar, V.V., Devan, R.S., Choudhary, R.J., Kedge, V.V., Pawar, N.V., Kim, J.H., Chougale, A.D. (2019). A green approach for the synthesis of α-Fe2O3 nanoparticles from *Gardenia resinifera* plant and it's In vitro hyperthermia application. *Heliyon* 5(7): 1–5.

Kaushik, N., Kaushik, N., Linh, N., Ghimire, B., Pengkit, A., Sornsakdanuphap, J., Lee, S.-J., Choi, E. (2019) Plasma and nanomaterials: Fabrication and biomedical applications. *Nanomaterials* 9(1): 98.

Khan, S.A., Ahmad, A. (2013) Fungus mediated synthesis of biomedically important cerium oxide nanoparticles. *Materials Research Bulletin* 48(10): 4134–4138.

Kouvaris, P., Delimitis, A., Zaspalis, V., Papadopoulos, D., Tsipas, S.A., Michailidis, N. (2012) Green synthesis and characterization of silver nanoparticles produced using *Arbutus unedo* leaf extract. *Materials Letters* 76: 18–20.

Kuppusamy, P., Yusoff, M.M., Maniam, G.P., Govindan, N. (2016) Biosynthesis of metallic nanoparticles using plant derivatives and their new avenues in pharmacological applications — An updated report. *Saudi Pharmaceutical Journal* 24(4): 473–484.

Lassenberger, A., Grünewald, T.A., Van Oostrum, P.D.J., Rennhofer, H., Amenitsch, H., Zirbs, R., Lichtenegger, H.C., Reimhult, E. (2017) Monodisperse iron oxide nanoparticles by thermal decomposition: Elucidating particle formation by second-resolved in situ small-angle X-ray scattering. *Chemistry of Materials* 29(10): 4511–4522.

Lee, H.J., Song, J.Y., Kim, B.S. (2013) Biological synthesis of copper nanoparticles using *Magnolia kobus* leaf extract and their antibacterial activity. *Journal of Chemical Technology and Biotechnology* 88(11): 1971–1977.

Li, W., Zhao, D. (2013) An overview of the synthesis of ordered mesoporous materials. *Chemical Communications* 49(10): 943–946.

Lohrasbi, S., Kouhbanani, M.A.J., Beheshtkhoo, N., Ghasemi, Y., Amani, A.M., Taghizadeh, S. (2019) Green synthesis of iron nanoparticles using *Plantago major* leaf extract and their application as a catalyst for the decolorization of azo dye. *BioNanoScience* 9(2): 317–322.

Maensiri, S., Laokul, P., Klinkaewnarong, J., Phokha, S., Promarak, V., Seraphin, S. (2008) Indium oxide (In_2O_3) nanoparticles using *Aloe vera* plant extract: Synthesis and optical properties. *Optoelectronics and Advanced Materials-Rapid Communications* 2(3): 161–165.

Makhlouf, A.S.H. (2011) 1—Current and advanced coating technologies for industrial applications. In: A.S.H. Makhlouf, I. Tiginyanu (Eds.) *Nanocoatings and Ultra-Thin Films* (pp. 3–23). Woodhead Publishing.

Meidanchi, A., Jafari, A. (2019) Synthesis and characterization of high purity Ta_2O_5 nanoparticles by laser ablation and its antibacterial properties. *Optics & Laser Technology* 111: 89–94.

Moy, A.J., Schwartz, J.M., Chen, R., Sadri, S., Lucas, E., Cato, K.D., Rossetti, S.C. (2021) Measurement of clinical documentation burden among physicians and nurses using electronic health records: A scoping review. *Journal of the American Medical Informatics Association* 28(5): 998–1008.

Naika, H.R., Lingaraju, K., Manjunath, K., Kumar, D., Nagaraju, G., Suresh, D., Nagabhushana, H. (2015) Green synthesis of CuO nanoparticles using *Gloriosa superba* L. extract and their antibacterial activity. *Journal of Taibah University for Science* 9(1): 7–12.

Nanotechnology. (n.d.). Retrieved January 5, 2023, from www.nano.gov/nanotech-101/what/ definition

Nava-Avendaño, J., Nussbaum, M., Veilleux, J. (2021) Thermal plasma synthesis of Li_2S nanoparticles for application in lithium-sulfur batteries. *Plasma Chemistry and Plasma Processing* 41(4): 1149–1167.

Órdenes-Aenishanslins, N.A., Saona, L.A., Durán-Toro, V.M., Monrás, J.P., Bravo, D.M., Pérez-Donoso, J.M. (2014) Use of titanium dioxide nanoparticles biosynthesized by *Bacillus mycoides* in quantum dot sensitized solar cells. *Microbial Cell Factories* 13(1): 1–10.

Padman, A.J., Henderson, J., Hodgson, S., Rahman, P.K.S.M. (2014) Biomediated synthesis of silver nanoparticles using *Exiguobacterium mexicanum. Biotechnology Letters* 36(10): 2079–2084.

Park, J.H., Ahn, H.S. (2020) Electrochemical synthesis of multimetallic nanoparticles and their application in alkaline oxygen reduction catalysis. *Applied Surface Science* 504: 144–517.

Pestman, J.M., Engberts, J.B.F.N., De Jong, F. (2010) Sonochemistry: Theory and applications. *Recueil Des Travaux Chimiques Des Pays-Bas* 113(12): 533–542.

Ramimoghadam, D., Bagheri, S., Hamid, S.B.A. (2014) Progress in electrochemical synthesis of magnetic iron oxide nanoparticles. *Journal of Magnetism and Magnetic Materials* 368: 207–229.

Ramrakhiani, L., Ghosh, S., Majumdar, S. (2016) Surface modification of naturally available biomass for enhancement of heavy metal removal efficiency, upscaling prospects, and management aspects of spent biosorbents: A review. *Applied Biochemistry and Biotechnology* 180(1): 41–78.

Razack, S.A., Suresh, A., Sriram, S., Ramakrishnan, G., Sadanandham, S., Veerasamy, M., Nagalamadaka, R.B., Sahadevan, R. (2020) Green synthesis of iron oxide nanoparticles using *Hibiscus rosa-sinensis* for fortifying wheat biscuits. *SN Applied Sciences* 2(5): 1–9.

Saima, B., Khan, N., Al-Faiyz, Y.S.S., Ludwig, R., Rehman, W., Habib-ur-Rehman, M., Sheikh, N.S., Ayub, K. (2019) Photo-tunable linear and nonlinear optical response of cyclophanediene-dihydropyrene photoswitches. *Journal of Molecular Graphics and Modelling* 88: 261–272.

Salem, D.M.S.A., Ismail, M.M., Aly-Eldeen, M.A. (2019) Biogenic synthesis and antimicrobial potency of iron oxide (Fe_3O_4) nanoparticles using algae harvested from the Mediterranean Sea, Egypt. *Egyptian Journal of Aquatic Research* 45(3): 197–204.

Satyanarayana, T. (2018) A review on chemical and physical synthesis methods of nanomaterials. *International Journal for Research in Applied Science and Engineering Technology* 6(1): 2885–2889.

Senthilkumar, R.P., Bhuvaneshwari, V., Ranjithkumar, R., Sathiyavimal, S., Malayaman, V., Chandarshekar, B. (2017) Synthesis, characterization and antibacterial activity of hybrid chitosan-cerium oxide nanoparticles: As bionanomaterials. *International Journal of Biological Macromolecules* 104: 1746–1752.

Singh, J., Singh, N., Rathi, A., Kukkar, D., Rawat, M. (2017) Facile approach to synthesize and characterization of silver nanoparticles by using mulberry leaves extract in aqueous medium and its application in antimicrobial activity. *Journal of Nanostructures* 7(2): 134–140.

Singh, P., Kim, Y.J., Wang, C., Mathiyalagan, R., Farh, M.E.-A., Yang, D.C. (2016) Biogenic silver and gold nanoparticles synthesized using red ginseng root extract, and their applications. *Artificial Cells, Nanomedicine, and Biotechnology* 44: 811–816.

Singaravelan, R., Bangaru Sudarsan Alwar, S. (2015) Electrochemical synthesis, characterisation and phytogenic properties of silver nanoparticles. *Applied Nanoscience* 5(8): 983–991.

Sportelli, M., Valentini, M., Picca, R., Milella, A., Nacci, A., Valentini, A., Cioffi, N. (2018). New insights in the ion beam sputtering deposition of ZnO-fluoropolymer nanocomposites. *Applied Sciences* 8(1): 77.

Suslick, K.S., Casadonte, D.J., Green, M.L.H., Thompson, M.E. (1987) Effects of high intensity ultrasound on inorganic solids. *Ultrasonics* 25(1): 56–59.

Tavares, J., Swanson, E.J., Coulombe, S. (2008) Plasma synthesis of coated metal nanoparticles with surface properties tailored for dispersion. *Plasma Processes and Polymers* 5(8): 759–769.

Tonelli, D., Scavetta, E., Gualandi, I. (2019) Electrochemical deposition of nanomaterials for electrochemical sensing. *Sensors* 19(5): 1186.

Vesel, A., Zaplotnik, R., Mozetič, M., Primc, G. (2021) Surface modification of PS polymer by oxygen-atom treatment from remote plasma: Initial kinetics of functional groups formation. *Applied Surface Science* 561: 150058.

Wang, T., Yang, L., Zhang, B., Liu, J. (2010) Extracellular biosynthesis and transformation of selenium nanoparticles and application in H_2O_2 biosensor. *Colloids and Surfaces B: Biointerfaces* 80(1): 94–102.

Yudasari, N., Wiguna, P.A., Handayani, W., Suliyanti, M.M., Imawan, C. (2021) The formation and antibacterial activity of Zn/ZnO nanoparticle produced in Pometia pinnata leaf extract solution using a laser ablation technique. *Applied Physics A* 127(1): 56.

Zeiger, B.W., Suslick, K.S. (2011) Sonofragmentation of molecular crystals. *Journal of the American Chemical Society* 133(37): 14530–14533.

Zhang, T., Doert, T., Wang, H., Zhang, S., Ruck, M. (2021) Inorganic synthesis based on reactions of ionic liquids and deep eutectic solvents. *Angewandte Chemie International Edition* 60(41): 22148–22165.

Zheng, D., Hu, C., Gan, T., Dang, X., Hu, S. (2010) Preparation and application of a novel vanillin sensor based on biosynthesis of Au-Ag alloy nanoparticles. *Sensors and Actuators, B: Chemical* 148(1): 247–252.

4 Bionanotechnology in Agriculture

Current Applications, Challenges, and Future Prospects

Esha Rami, Richa Das, Shreni Agrawal, and Vijay Upadhye

4.1 INTRODUCTION

The agricultural sector is confronted with a number of difficulties, including non-sustainable agricultural methods, urbanization, industrialization, and climate change. Over the next 30 years, the growth in the world's population is anticipated to cause demand to rise exponentially, exacerbating this issue even further. Agricultural products are now seen as raw materials for various forms of commercial production due to the depletion of natural energy supplies [1]. Critical problems include a lack of productive land, high fertilizer and pesticide costs, a shortage of cropland, unemployment, and malnourishment in emerging nations where agriculture is a major contributing factor to GDP. As a result, developing technologies to solve these difficulties for food security is necessary in addition to changes in the governance and rules surrounding these issues [2]. Due to rising population, climatic variability, industrial pollution, and increased water and energy shortages, the manufacturing and distribution of food worldwide is under tremendous strain. Currently, agriculture consumes an incredible amount of resources. For instance, 2.7 trillion cubic meters of water (approximately 70% of all freshwater consumed globally) and 187 million tonnes of fertilizer and pesticides are needed to produce three billion tonnes of crops annually, also consuming more than two quadrillion British thermal units (BTU) of energy [3]. By 2050, the world's population is expected to exceed 10 billion, which will significantly increase food demand, particularly in emerging nations. In addition, it is estimated that by 2050, 2 billion additional individuals will fall into the category of being malnourished, which today numbers 815 million people. The international agricultural systems must undergo significant modifications as a result of this circumstance. The efficacy of agricultural inputs can be increased through nanotechnology, which has the potential to significantly improve the agriculture industry [4].

Nanobiotechnology combines biology with nanotechnology. The most recent development in the multidisciplinary subject of nanobiotechnology is the fusion of

DOI: 10.1201/9781003362258-4

biological and nonbiological components with living organisms through the use of nanotechnology. A biophysicist from Cornell University in the United States named Lynn W. Jelinski coined the term "nanobiotechnology," which describes scales between 1 and 100 nm. Genetic engineering and breeding initiatives use nanobiotechnology [5]. Numerous substances, including lipids, polymers, emulsions, semiconductor quantum dots (QDs), silicates, magnetic compounds, and metal oxides, can be used to create nanoparticles (NPs) [6–9].

There are many potential uses for nanotechnology in climate-resilient and intelligent agriculture. Nanostructures can minimize nutrient deficiencies during fertilization, decrease the number of chemical products that need to be sprayed, and boost productivity through improved water and nutrient administration [10]. The use of nanoparticles to control plant diseases, insect pests, and weeds has a lot of potential to be both environmentally friendly and economically viable [11–14]. For instance, using nano-silver compounds instead of commercial fungicides effectively controlled various fungal pathogens [15]. Unique photoelectric, physiological, and catalytic capabilities of NPs help plants grow faster, produce more photosynthesis, and withstand abiotic and biotic stress. For instance, due to their numerous surface oxygen vacancies that fluctuate between two oxidation states (Ce^{3+} and Ce^{4+}), nanoscale CeO_2 particles are efficient scavengers of reactive oxygen species (ROS). The ability of plants to respond to stress and ultimately survive under pressure can be enhanced by this antioxidant-enzyme-mimicking action [16,17]. Additionally, nanosensors have shown promise in the field of agriculture. Nanosensors are used in agriculture to assess crop pest identification, nutritional requirements, soil moisture, and pesticide residue. The high sensitivity and low detection limit of nanosensors make them more advantageous for intelligent agriculture. The development of nanosensors for pathogen detection has made use of numerous metal nanomaterials, such as gold nanoparticles (Au NP), carbon nanotubes (CNTs), quantum dots (QDs), and different nanocomposites with polymers, as shown in Figure 4.1 [18–21].

4.2   PLANT GENETIC ENGINEERING WITH NANOMATERIALS

One of the most promising new areas of scientific and technical development is nanobiotechnology, which allows for the creation and molecular manipulation of biological and biochemical materials, equipment, and systems. Comparing the standard transformation process to one mediated by nanoparticles has various benefits. Traditional transformation techniques have some limitations, such as the fact that some procedures work better with monocotyledonous plants than they do with dicotyledonous ones [22]. However, NP-based transformations apply to both monocotyledonous and dicotyledonous plants. Transgenic gene silencing is also overcome by nanoparticle-mediated transformation [23]. Gene transformation mediated by NPs is more effective than conventional transformation. Additionally, a variety of NPs, including as ceramics, silicates, metal oxides, magnetic materials, dendrimers, and liposomes, can transport multiple genes without transformation or genomic barriers [24], and mesoporous silica, are used for transformation [25].

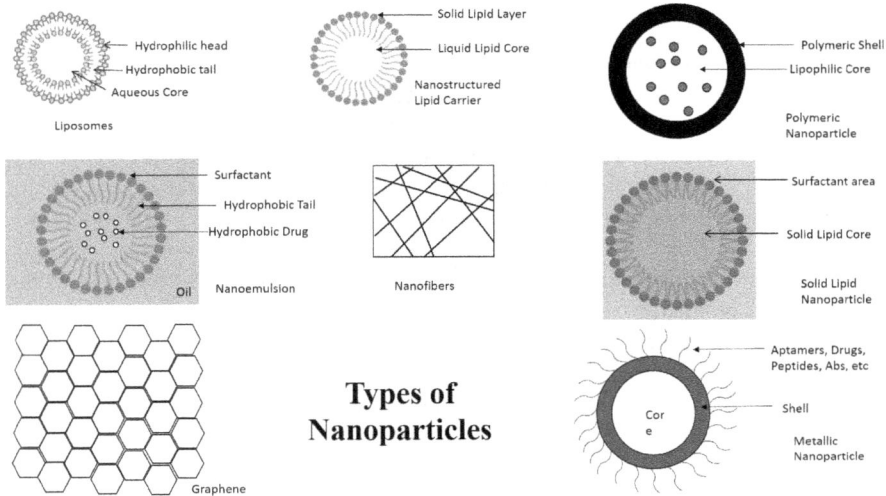

FIGURE 4.1 Different types of nanomaterials used in agriculture development.

For a successful genetic transformation, nanomaterials must be less cytotoxic and genotoxic, biocompatible, and biodegradable. About 80.7% of the transition is facilitated by NPs, but the capability of the *Agrobacterium tumefaciens*-mediated transformation is just 54.4%. Only 8% of bare DNA is utilized in conversions facilitated by NPs. These encouraging findings imply that NPs may be more effective plant transformation agents than *Agrobacterium tumefaciens* [26]. Numerous crops, including cotton, maize, rice, soybean, and tobacco, have been observed to undergo a transformation that is mediated by silicon carbide-based nanoparticles [27,28].

Despite their enormous significance, some obstacles are preventing NPs from being used effectively in Genetic Engineering (GE). Numerous studies have shown that plants' absorption of NPs causes an obstruction in their vascular system, which results in structural damage to the plant's DNA and causes oxidative stress [29–31]. Some NPs' high oxidative properties affect plant cells' normal metabolism and interfere with their ability to be genetically controlled, leading to oxidative burst of the transformed cells [31,32]. Furthermore, the strong biomolecule binding to NPs and the dissolution of the binding complex in plant cells render NP-mediated GE useless [33,34]. Because they have different binding affinities with other NPs based on their structure, charge, chemical composition, and surface area, many biomolecules are excellent candidates for a bioconjugation complex [35,36].

Researchers have tried to comprehend how an NP-biomolecule bioconjugated complex will be provided in a force-independent manner for the potential future of NP-mediated GE [37]. These nanocarriers have been studied for their capacity to transport particular compounds to plant organelles while avoiding injury to the transformed cells, leaving minimal residue on the offspring cells, and ensuring no adverse impact on the plant or the environment [38].

4.3 NANOFERTILIZERS

Chemical substances known as fertilizers provide plants with nutrients [39]. Since the early 1950s, chemical fertilizers have been used. However, overuse of chemical fertilizers has reduced soil fertility, intensified eutrophication, and worsened the issue of water contamination. The use of nanofertilizers has the potential to significantly reduce the environmental issues brought on by the use of chemical fertilizers. Because they contain nutrients and growth-promoting compounds found in nanopolymers, chelates, or emulsions, nanofertilizers can improve plant growth [40,41]. The application of iron-chelated nanofertilizer at 4 kg ha^{-1} on Varamin 88 and Viroflayand spinach types, resulted in improvements of 58% and 47%, respectively, in wet weight as well as improved leaf surface index and aerial organs [42]. The amounts of Zn, N, and P absorption in coffee leaves increased after $ZnSO_4$ was loaded onto a chitosan NP emulsion to create a zinc-boron nanofertilizer [43]. Starch accumulation in root tips was accelerated by the application of chitosan-polymethacrylic acid (PMAA) NP to pea plants. Additionally, the production of proteins including convicilin, vicilin, and legumin increased [44] (Figure 4.2)

Nanotechnology in agriculture has the potential to significantly improve organic waste decomposition and compost generation. However, the study is still in its early phases and there have not been any clear findings revealed yet. The information that is currently available makes it clear that nanofertilizers, because of their high use efficiency, can lower the amounts of fertilizers that need to be applied, reducing the environmental effects caused by nutrient losses. This breakthrough comes just in time

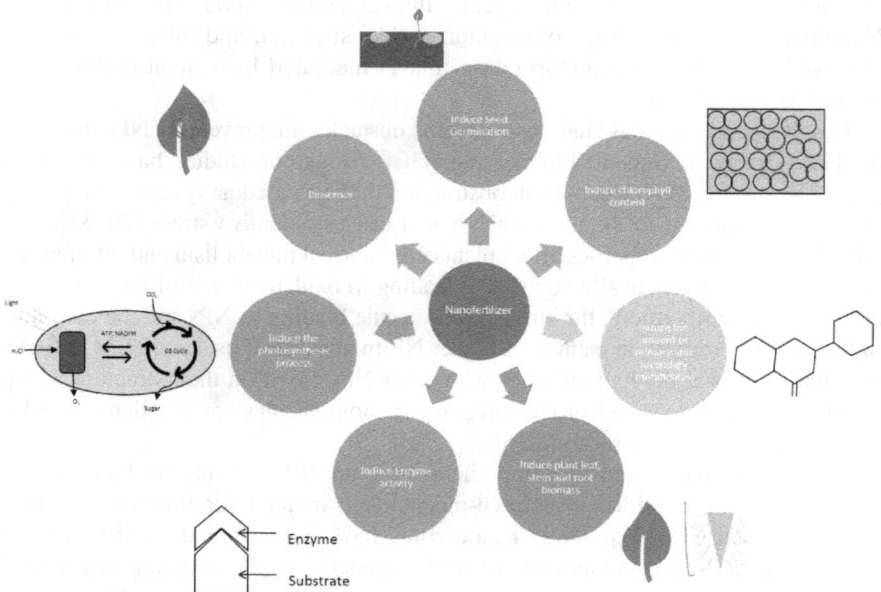

FIGURE 4.2 Effects of nanofertilizers on plant production for agriculture development.

to protect the environment and assure global food security. However, research on the economic viability of nanofertilizers is still required for productive and sustainable agriculture [4].

4.4 NANOPESTICIDES

The balance of the ecosystem is impacted by the indiscriminate and illogical use of pesticides, which also puts everyone's health at risk. Pesticide residues in food and water can be consumed accidentally or on the job, and can have both short- and long-term deadly consequences [45]. Chemical stability, solubility, bioavailability, photodecomposition, and soil absorption all affect pesticide toxicity [46]. By creating nanocarriers that allow for the delayed release of pesticides, nanotechnology's primary objective is to decrease these effects [47].

Polymer nanoparticles are used to create the pesticide formulations and encapsulation methods that are most desirable [46]. Pesticides that are very dangerous to the environment are degraded using photocatalytic materials [48]. TiO_2 NPs completely removed pesticides including chlorpyrifos, cypermethrin, and chlorothalonil after being exposed to UVA light for 30 minutes [49]. Meanwhile, Cu-doped ZnO was used to study the breakdown of the insecticide monocrotophos. An electron can move from the valence band to the conduction band when Cu generates an intermediate band. This increased optical absorption narrows the bandgap and shows a significant decline in the effectiveness of monocrotophos pesticides [50]. The benefits of applying specific nanostructured compounds to plants include enhanced yield, durability, and nutrient content. Although these nano-compounds can protect against pests (insects, plants, and bacteria) and provide nutrients, they can also cause stress to other ecosystem species, posing harm to the wider environment [51].

A severe risk to human exposure is posed by dangerous nanopesticides that can pass through biological barriers such as the blood–brain barrier, blood–placental barrier, and blood–retinal barrier, and could cause long-term injury to organ systems [45]. Nanopesticides from agricultural and industrial wastewater runoff infiltrate the water supply when soil leaching happens during a precipitation event, altering its quality, extending human exposure time, and creating problems for the ecosystem. The strong capacity of nanoparticles for diffusion and bioaccumulation in soil, aquatic habitats, foods, and animals, has been noted as potential sources of toxicological consequences [52–55]. Humans may experience a variety of adverse consequences linked to susceptibility and nanoparticle exposure time, including acute and chronic pathological symptoms that affect the respiratory, cardiovascular, lymphatic, autoimmune, neurological, and other systems. Due to bioaccumulation, these manifestations may appear immediately upon exposure or years later [56].

4.5 SOIL–PLANT SYSTEMS USING NANOBIOSENSORS

Nanobiosensors provide many benefits over conventional and earlier-generation sensors, including a high surface-to-volume ratio, quick electron-transfer kinetics, high sensitivity, stability, and extended life. With the aid of nanobiosensors, agricultural yields can be quickly boosted by managing water, soil, fertilizers, and pesticides properly [57]. Nano-structured particles are employed as nanobiosensors with enhanced characteristics of rapid reaction, sensitivity, stability, and potential for repeated application. Graphene oxide sheets and Ag-Pd interdigitated electrodes are two examples of nanobiosensors that have been treated with ceramic substrates and generated with a range of sensitivity and response levels. These sensors take advantage of the characteristics of NMs and ceramic materials by using the ion transport and dispersion concentrations of graphene oxide sheets [58,59]. For measuring the total carbon, organic matter, sodium chloride, phosphate, and residual nitrate in the soil, NM-based biosensors are still in their infancy [60]. The most extensively used fertilizer for agricultural production is urea, which is also a key cause of the water pollutants which causes eutrophication and have negative effects on the environment. These pollutants are found in soil and water, and nanosensors are utilized to detect them via microfluidic impedimetric and colorimetric testing [61,62].

Smart farming's lower hazards, the widespread use of food and agricultural goods derived from nanomaterials, and the less likely immobilization of nanosensors have all generated questions about the environment and human health (Table 4.1). The monitoring of nano bio–eco interactions in soils is hampered by their complexity. A thorough approach is necessary to comprehend these interactions in the soil, plants, air, and eventually in the food chain. To identify, validate, and lessen their hazardous effects on the entire environment, nanomaterials should be used sustainably. Regulatory bodies and legislation can provide the roadmaps and directions for this [4].

4.6   FUTURE PROSPECTS

The potential applications of nanoparticles in agriculture, including their novel method of applying fertilizers and pesticides, their use in plant genetic engineering, and their use in smart farming with nanobiosensors, may eventually result in the development of environmentally friendly and sustainable agricultural technology [87]. Nanomaterials in food and agriculture are, however, raising new safety concerns [88]. Due to emerging technologies, the environment is being exposed to an increasing amount of manmade nanomaterials. The use of nanocarriers in agriculture is currently restricted by production scale and price. The successful application of nanomaterials in agriculture and their large-scale production will result in a significant reduction in costs. The commercialization of nanomaterials for agricultural applications is difficult because it necessitates highly protected materials, superior testing priority, accurate risk assessments, and international regulatory guidelines [89]. Immediately after nanoparticles are introduced into the agro-environment, several changes take place. One of the main concerns is the unanticipated effects of nanoparticles on the human body. Nanoparticle waste may be harmful [90,91], and it can cause oxidative damage and unwanted reactions.

High levels of free radical production lead to oxidative stress in cells, which prevents them from carrying out regular redox-regulated biological tasks, which is how nanotoxicity is mediated [92]. Alveolar basal epithelial cell cancer can manifest

TABLE 4.1
Applications of Different Nanomaterials in the Field of Agriculture

S. no.	Nanomaterials	Crops	Concentration	Application	References
1	Multiwalled carbon nanotubes (MWCNTs)	*Eysenhardtia polystachya*	Doses of 20 μg/ml and 40 μg/mL	Induces seed germination within a few days, significantly promoted leaf number, root growth, and the dry and fresh weights of shoots and roots of seedlings	[63]
2	Carbon nanotubes (CNTs)	Tomato, onion, turnip, radish	Four concentrations of CNTs (0, 10, 20, and 40 mg L^{-1})	Improves seed germination in tomato and onion; The dry weight of tomato and radish shoots increased at all concentrations of CNTs	[64]
3	Multiwalled carbon nanotubes (MWCNTs). MWCNTs functionalized with carboxylic acids (MWCNT–COOH) and graphene	Bog birch (*Betula pumila* L.) and Labrador tea (*Rhododendron groenlandicum* L.)	20 or 40 μg/mL CNT	Nanopriming of seeds with CNTs enhances seed germination, propagation, and seedling vigor in non-resource boreal forest species	[65]
4	Halloysite nanotubes (HNTs)	wheat seeds	Doses of 0.1, 1, and 10 mg mL^{-1}	Enhance biomass and chlorophyll content as well as significantly used in polymer composites, drug delivery, waste treatment, and cosmetics due to their special structure and biocompatibility	[66]
5	Carbon nanotubes (CNTs)	Broccoli (*Brassica oleracea* var. Italica)	Doses of 3 g/L, 5 g/L, and 7 g/L	CNTs with the combination of BA stimulate cell division, nucleic acid metabolism, root–shoot interactions under stress. CNTs with plant hormones also stimulate seed germination, mobilization of nutrients, break apical dominance, induce parthenocarpy, and indue flowering in plants	[67]

(Continued)

TABLE 4.1 (Continued)
Applications of Different Nanomaterials in the Field of Agriculture

S. no.	Nanomaterials	Crops	Concentration	Application	References
6	Phenylalanine-functionalized carbon nanotubes (f-MWCNTs) and pristine MWCNTs	Basil (*Ocimum basilicum* L.)	Doses of 0, 50, 100, and 200 mg/L	Induce callus formation and dry matter content (DMC) as well as induce the activity of catalase, polyphenol oxidase (PPO), peroxidase (POD), and L-phenylalanine ammonia-lyase. Enhance content of total phenolics, flavonoids, and individual phenolic acids	[68]
7	Water-soluble carbon nanoparticles (CNPs)	Lettuce (*Lactuca sativa*)	0.3% soluble carbon nanoparticles	Substantially enhance lettuce seed germination and post-germination growth under saline conditions	[69]
8	Multiwalled carbon nanotubes (MWCNTs); single-walled carbon nanotubes (SWCNTs); graphene	Tomato (*Solanum lycopersicum* L.)	5 mg/mL SWCNTs; concentrations of MWCNTs (0.05, 0.5, 5, and 50 mg/mL)	MWCNTs induce the production of nitric oxide via nitric reductase enzyme which causes the induction of lateral root formation	[70]
9	Multiwalled carbon nanotubes (MWCNTs)	Two maize varieties: Yuebaitiannuo7 and Yuecainuo2	Doses of 0, 100, and 200 mg L^{-1}	Enhanced root and shoot fresh weight and antioxidant enzyme activities, including peroxidase (POD), superoxide dismutase (SOD), and catalase (CAT) activities, and reduced the malonaldehyde (MDA) content under cadmium stress	[71]
10	Carbon nanotubes (CNTs)	Tomato, rapeseed, cucumber, and maize	100 mg CNT/kg of dry soil	CNT presence in soil causes positive impacts on cucumber and rapeseed (more than 50% increase in leaf biomass and surface area and 29% increase in chlorophyll for cucumber)	[72]
11	Pristine multiwalled-CNTs (MWCNTs)	Maize (*Zea mays*)	Doses of 0, 5, 10, 20, 40, 60 mg/L	Improve water absorption, plant biomass, and the concentrations of the essential Ca, Fe nutrients	[73]

12	Cu-chitosan nanoparticles	Maize (*Zea mays* L.)	0.01–0.16% concentration	Act as strong antifungal and antioxidant agent, enhance enzyme activities, increase plant height, stem diameter, root length, root number, and crop yield	[74]
13	Amine-modified polystyrene nanospheres, titanium dioxide (TiO_2) nanoparticles, sulfate-modified polystyrene nanospheres	Buttercrunch lettuce plants	Nano TiO_2 concentration: 100 µg/mL, sulfate and amine-modified nanospheres concentration: 50 µg/ml	Act as potent nanopesticides and nanofertilizers which eliminate the growth of microorganisms and also enhance seed germination and plant growth	[75]
14	Iron (III) oxide nanomaterials	Wheat (*Triticum aestivum* L.)		Fe_2O_3 NMs enhanced root length, plant height, biomass, and chlorophyll content of wheat. Fe_2O_3 NMs with 20–40 nm size could be proposed as a nanofertilizer for agricultural applications	[76]
15	Zinc oxide nanoparticles (ZnO NPs)	Cotton (*Gossypium hirsutum* L.)	Range of concentrations (25–200 mg L^{-1} ZnO NPs)	Significantly cause increases in the level of total biomass, chlorophyll a and b, carotenoids, total soluble protein contents, superoxide dismutase (SOD), and peroxidase (POX)	[77]
16	Metal oxide nanomaterials (ZnO and TiO_2 nanomaterial)	Solanaceae crops (eggplant, pepper, and tomato crops)	Four concentrations: 0, 50, 100, and 150 mg/L	ZnO and TiO_2 NPs promoted drought resistance of wheat cultivars at seedling stage via stabilizing the photosynthetic pigments, enhancing leaf water content and germination rate in plants, increased the antioxidative enzyme activity and slowed down lipid peroxidation	[78]

(*Continued*)

TABLE 4.1 (Continued)
Applications of Different Nanomaterials in the Field of Agriculture

S. no.	Nanomaterials	Crops	Concentration	Application	References
17	Metal (Zn and Cu)-based chitosan nanomaterials			Showed effective antimicrobial, antibacterial and antifungal activity and induced plant growth and biophysical characteristics of plant and boosted plant immune response. Also provided nutrition and helped in vigorous growth of plant for further protection from abiotic and biotic stresses. Induced the amylase and protease enzyme and some defense enzymes of plants which protect them from disease	[79]
18	Zerovalent iron (nZVI) nanoparticles	*Arabidopsis thaliana*	500 mg nZVI/kg	Induced accumulation of glucose, sucrose, and starch by enhancing the photosynthesis process, and photosynthetic-related inorganic nutrients such as phosphorus, Mn, and Zn were also increased.	[80]
19	Zinc oxide nanoparticles	Wheat (*Triticum aestivum* L.)	NP levels (0, 15, 62, 125, 250, and 500 mg/L)	Green ZnO nanoparticles can be used as nanofertilizers and nanopesticides. Also increase the germination rate of wheat seeds and increase the root and shoot length of plants	[81]
20	γ-Fe$_2$O$_3$ nanoparticles	*Citrus maxima*	20–100 mg/L γ-Fe$_2$O$_3$ NPs	No oxidative stress occurred under all Fe treatments. γ-Fe$_2$O$_3$ NPs reduce nutrient loss due to their strong adsorption ability. They can be used in medical diagnostics, controlled drug release, separation technologies, and environmental engineering	[82]

21	Cu–Zn micronutrient-carrying carbon nanofibers (CNFs)	Chickpea (*Cicer arietinum*)	Different doses PBMC nanofertilizer: 0.25, 0.50, 1.0, 2.0, and 4.0 g	CNFs enhance water uptake capacity and germination rate, subsequently enhancing growth of the plants. CNFs also have translocation ability within plants, thereby serving as a carrier for the micronutrients within the plant	[83]
22	TiO_2 nanoparticles	Wheat (*Triticum aestivum* L.)	10 mg/L, 100 mg/L, and 1000 mg/L, and a control without TiO_2 NPs	Elevated level of CO_2 under all the TiO_2 nanoparticle concentrations increases root biomass and large numbers of lateral roots	[84]
23	Carbon nanoparticles (CNPs)	*Vigna radiata* (L.)	Concentrations of 25–200 µM	Increase in biomass, total chlorophyll, and protein content in *V. radiata* seedlings. They promote nutrient absorption and accumulation amount and also contribute to pollutant removal and soil remediation; hence they are a preferable choice as a nanofertilizer	[85]
24	Nano silicon dioxide	Maize		Drought resistance, increment in lateral root roots number along with shoot length	[86]
25	Colloidal silica + NPK fertilizers	Tomato		Increased resistance to pathogens	[86]

in people when the nano-clay in low-density polyethylene clumps breaks down [93]. Numerous investigations have focused on Ag's antibacterial mechanism. However, as it is a well-known heavy metal, denaturing proteins and enzymes in large quantities can increase its toxicity in the body [94]. Due to their insoluble nature, TiO_2, Ag, and carbon nanotubes have all been shown through research to be able to enter the bloodstream and accumulate in organs [95]. TiO_2 can cause genotoxicity, which results in chromosomal instability, and oxidative stress, which causes inflammation when it is used as a food additive [96].

The development of mechanisms that would improve herbicides' release profiles without affecting their characteristics and the creation of innovative carriers with improved action are two important topics that need additional research. The diverse uses of nanomaterials, including their manufacturing, toxicity, and utilization at the field level, still require more study [87]. Plant gene transformation may be hampered by the physical and chemical characteristics of nanoparticles, such as size, inertness, charge, composition, and low water dispersibility [97]. The stability of nanoparticle colloids plays a crucial role in gene transformation. The ionic strength, osmotic concentration, and other components of the buffer affect how colloidally stable nanoparticles are. Additionally, the nanoparticles may not be compatible with all plant cell culture buffers due to their instability in both the individually distributed and colloidal state. Nanoparticle surface modification techniques including charge, elasticity, and synthesis may become more widely used as a result of recent advancements [97,98]. Additionally, specific DNA/RNA characteristics must be adjusted to transfer the DNA/RNA to a range of tissues, such as pollen, calluses, embryonic tissue, and germ-line cells. These characteristics include buffer conditions, DNA/nanoparticle ratio, exposure length, and sterility [99]. The parameters of the payloads (DNA/RNA), including buffer conditions, DNA/nanoparticle ratio, exposure time, and sterility, must also be altered for distribution to different tissues (e.g. pollen, callus, embryonic tissue, and germ-line cells).

4.7 CONCLUSION

This review chapter integrates the use of nanotechnology in the field of agriculture to minimize the effects of climate change, chemical pesticides, and soil erosion to plant growth and its quality and safety. Nanomaterials when combined with plant nutrients lead to the development of nanofertilizers. These nanofertilizers improve the soil quality, plant biomass, nutrient absorption rate of lateral roots, germination rate of seedlings, chlorophyll content in leaves, photosynthesis ability, translocation ability of nutrients in plant, root and stem diameter, and production of secondary metabolites such as phenolics, flavonoids, and carotenoids, etc. Additionally, these nanofertilizers can reduce biotic and abiotic stresses. The application of nanoscale sensor mechanisms, sometimes known as nanobiosensors, is another aspect of nanotechnology. With proper management of water, soil, fertilizers, and pesticides, agricultural yields can be increased quickly and with early detection. Nano-structured particles are employed as nanobiosensors with enhanced characteristics of rapid reaction, sensitivity, stability, and potential for repeated application. Several nanopesticides have also been used as antimicrobial agents which inhibit the growth of microbes on plants and improve their sustainability. The use of such nanoproducts involves the use of different nanoparticles which sometimes leads to toxic effects on plant growth. It can cause abnormal cell division, aggregation of NPs on the surface of plants, plant morphology, etc. However, by resolving such problems, nanoproducts in agriculture could be used widely in the future as they improve the growth of plants and their quality as compared to other methods, more efficiently.

REFERENCES

1. Gruère, G., Narrod, C., Abbott, L. (2011) *Agricultural, Food, and Water Nanotechnologies for the Poor*. International Food Policy Research Institute, Washington, DC.
2. Biswal, S.K., Nayak, A.K., Parida, U.K., Nayak, P. (2012) Applications of nanotechnology in agriculture and food sciences. *International Science Innovation Discovery* 2: 21–36.
3. Kah, M., Tufenkji, N., White, J.C. (2019) Nano-enabled strategies to enhance crop nutrition and protection. *Nature Nanotechnology* 14, 532–540.
4. Usman, M., Farooq, M., Wakeel, A., Nawaz, A., Cheema, S. A., Rehman, H. ur, … Sanaullah, M. (2020) Nanotechnology in agriculture: Current status, challenges and future opportunities. *Science of The Total Environment* 137778. doi:10.1016/j.scitotenv.2020.1377
5. Scrinis, G., Lyons, K. (2007) The emerging nano-corporate paradigm: nanotechnology and the transformation of nature, food and agri-food systems. *International Journal of the Sociology of Agriculture and Food* 15: 22–44.
6. Niemeyer, C.M., Doz, P. (2001) Nanoparticles, proteins, and nucleic acids: biotechnology meets materials science. *Angewandte Chemie International Edition* 40: 4128–4158.
7. Oskam, G. (2006) Metal oxide nanoparticles: synthesis, characterization and application. *Journal of Sol-Gel Science and Technology* 37: 161–164.
8. Puoci, F., Lemma, F., Spizzirri, U.G., Cirillo, G., Curcio, M., Picci, N. (2008) Polymer in agriculture: a review. *American Journal of Agriultural and Biological Sciences* 3: 299–314.
9. Prasad, R., Bhattacharyya, A., Nguyen, Q.D. (2017) Nanotechnology in sustainable agriculture: Recent developments, challenges, and perspectives. *Frontiers in Microbiology* 8: 1014.
10. Gogos, A., Knauer, K., Bucheli, T.D. (2012) Nanomaterials in plant protection and fertilization: current state, foreseen applications, and research priorities. *Journal of Agricultural and Food Chemistry* 60(39): 9781–9792.
11. Ahmed, S., Ahmad, M., Swami, B.L., Ikram, S. (2016) A review on plant extract mediated synthesis of silver nanoparticles for antimicrobial applications: a green expertise. *Journal of Advanced Research* 7(1): 17–28.
12. Suman, P.R., Jain, V.K., Varma, A. (2010) Role of nanomaterials in symbiotic fungus growth enhancement. *Current Science* 99: 1189–1191
13. Chhipa, H., Joshi, P. (2016) Nanofertilisers, nanopesticides and nanosensors in agriculture. In: Ranjan, S., Dasgupta, N., Lichtfouse, E. (Eds.) *Nanoscience in Food and Agriculture*, vol 1. pp 247–282. Sustainable Agriculture Reviews. Springer, Cham.
14. Pandey, S., Giri, K., Kumar, R., Mishra, G., Rishi, R.R. (2016) Nanopesticides: opportunities in crop protection and associated environmental risks. *Proceedings of the National Academy of Sciences, India, Section B: Biological Sciences* 2016: 1–22.

15. Ouda, S.M. (2014) Antifungal activity of silver and copper nanoparticles on two plant pathogens, *Alternaria alternata* and *Botrytis cinerea*. *Research in Microbiology* 9(1): 34–42.

16. Dutta, P., Pal, S., Seehra, M.S., Shi, Y., Eyring, E.M., Ernst, R.D. (2006) Concentration of Ce^{3+} and oxygen vacancies in cerium oxide nanoparticles. *Chemistry of Materials* 18(21): 5144–5146.

17. Walkey, C., Das, S., Seal, S., Erlichman, J., Heckman, K., Ghibelli, L., Traversa, E., McGinnis, J.F., Self, W.T. (2015) Catalytic properties and biomedical applications of cerium oxide nanoparticles. *Environmental Science: Nano* 2(1); 33–53.

18. Zheng, Z., Li, X., Dai, Z., Liu, S., Tang, Z. (2011) Detection of mixed organophosphorus pesticides in real samples using quantum dots/bi-enzyme assembly multilayers. *Journal of Materials Chemistry* 21(42); 16955.

19. Cesarino, I., Moraes, F.C., Lanza, M.R.V., Machado, S.A.S. (2012) Electrochemical detection of carbamate pesticides in fruit and vegetables with a biosensor based on acetylcholinesterase immobilised on a composite of polyaniline–carbon nanotubes. *Food Chemistry* 135(3): 873–879.

20. Liu, L., Chen, W., Wei, J., Li, X., Wang, Z., Jiang, X. (2012) *Analytical Chemistry* 84(9): 4185–4191.

21. Talarico, D., Arduini, F., Amine, A. et al. (2016) Screen-printed electrode modified with carbon black and chitosan: a novel platform for acetylcholinesterase biosensor development. *Analytical and Bioanalytical Chemistry* 408: 7299–7309.

22. Frame, B.R., Shou, H., Chikwamba, R.K. et al. (2002) *Agrobacterium tumefaciens*-mediated transformation of maize embryos using a standard binary vector system. *Plant Physiology* 129: 13–22.

23. Malik, M.A., Athar, A., Saba, S. et al. (2017) Methods in transgenic technology. *Plant Biotechnology Principles and Applications* 279: 93–115.

24. Patra, J.K., Das, G., Fraceto, L.F. et al. (2018) Nano based drug delivery systems: recent developments and future prospects. *Journal of Nanobiotechnology* 16: 71–71.

25. Singh, A., Rajput, V., Singh, A.K. et al. (2021) Transformation techniques and their role in crop improvements: a global scenario of GM crops. In: *Policy Issues in Genetically Modified Crops*, P. Singh, A. Borthakur, A.A. Singh, et al. (Eds.), pp. 515–542. Academic Press.

26. Naqvi, S., Maitra, A.N., Abdin, M.Z. et al. (2012) Calcium phosphate nanoparticle mediated genetic transformation in plants. *Journal of Materials Chemistry* 22: 3500–3507.

27. Arshad, M., Zafar, Y., Asad, S. (2013) Silicon carbide whisker-mediated transformation of cotton (*Gossypium hirsutum* L.). In: *Transgenic Cotton: Methods and Protocols*, B. Zhang (Ed.), pp. 79–92. Totowa, NJ: Humana Press.

28. Lau, H.Y., Wu, H., Wee, E.J.H. et al. (2017) Specific and sensitive isothermal electrochemical biosensor for plant pathogen DNA detection with colloidal gold nanoparticles as probes. *Scientific Reports* 7: 38896–38896.

29. Pachapur, V.L., Dalila Larios, A., Cledón, M., Brar, S.K., Verma, M., Surampalli, R.Y., et al. (2016) Interaction of metal oxide nanoparticles with higher terrestrial plants: physiological and biochemical aspects. *Plant Physiology and Biochemistry* 110: 210–225.

30. Rastogi, A., Zivcak, M., Sytar, O., Kalaji, H.M., He, X., Mbarki, S., et al. (2017) Impact of metal and metal oxide nanoparticles on plant: a critical review. *Frontiers in Chemistry* 5: 78.

31. Du, W., Tan, W., Peralta-Videa, J.R., Gardea-Torresdey, J.L., Ji, R., Yin, Y., et al. (2017) Interaction of metal oxide nanoparticles with higher terrestrial plants: physiological and biochemical aspects. *Plant Physiology and Biochemistry* 110: 210–225.
32. Hossain, Z., Mustafa, G., Komatsu, S. (2015) Plant responses to nanoparticle stress. *International Journal of Molecular Science* 16: 26644–26653.
33. Saptarshi, S.R., Duschl, A., Lopata, A.L. (2013) Interaction of nanoparticles with proteins: relation to bio-reactivity of the nanoparticle. *Journal of Nanobiotechnology* 11: 26.
34. Fleischer, C.C., Payne, C.K. (2014) Nanoparticle-cell interactions: molecular structure of the protein corona and cellular outcomes. *Accounts of Chemical Research* 47: 2651–2659.
35. Nel, A.E., Mädler, L., Velegol, D., Xia, T., Hoek, E.M.V., Somasundaran, P., et al. (2009) Understanding biophysicochemical interactions at the nano-bio interface. *Nature Materials* 8: 543–557.
36. Dasgupta, S., Auth, T., Gompper, G. (2014) Shape and orientation matter for the cellular uptake of nonspherical particles. *Nano Letters* 14: 687–693.
37. Busch, R.T., Karim, F., Weis, J., Sun, Y., Zhao, C., Vasquez, E.S. (2019) Optimization and structural stability of gold nanoparticle-antibody bioconjugates. *ACS Omega* 4: 15269–15279.
38. Hu, P., An, J., Faulkner, M.M., Wu, H., Li, Z., Tian, X., et al. (2020) Nanoparticle charge and size control foliar delivery efficiency to plant cells and organelles. *ACS Nano* 14: 7970–7986.
39. Bottoms, M., Emerson, S.H. (2013) Chemistry, fertilizer, and the environment. *California Foundation for Agriculture in the Classroom* 1–98.
40. Naderi, M.R., Danesh-Shahraki, A. (2013) Nanofertilizers and their roles in sustainable agriculture. *International Journal of Agriculture and Crop Sciences* 5: 2229–2232.
41. Mehrazar, E., Rahaie, M., Rahaie, S. (2015) Application of nanoparticles for pesticides, herbicides, fertilisers and animals feed management. *International Journal of Nanoparticles* 8: 1–19.
42. Moghadam, A., Vattani, H., Baghaei, N., Keshavarz, N. (2012) Effect of different levels of fertilizer nano-iron chelates on growth and yield characteristics of two varieties of spinach (*Spinacia oleracea* L.): Varamin 88 and Viroflay. *Research Journal of Applied Sciences, Engineering and Technology* 4: 4813–4818.
43. Wang, S.L., Nguyen, A.D. (2018) Effects of Zn/B nanofertilizer on biophysical characteristics and growth of coffee seedlings in a greenhouse. *Research on Chemistry Intermediates* 44: 4889–4901.
44. Khalifa, N.S., Hasaneen, M.N. (2018) The effect of chitosan–PMAA–NPK nanofertilizer on *Pisum sativum* plants. *3Biotech* 8: 193–205.
45. Chaud, M., Souto, E.B., Zielinska, A., Severino, P., Batain, F., Oliveira-Junior, J., Alves, T. (2021) Nanopesticides in agriculture: Benefits and challenge in agricultural productivity, toxicological risks to human health and environment. *Toxics* 9(6): 131.
46. Kumar, S., Nehra, M., Dilbaghi, N., Marrazza, G., Hassan, A.A., Kim, K.H. (2019) Nano-based smart pesticide formulations: Emerging opportunities for agriculture. *Journal of Controlled Release* 294: 131–153.
47. Jampílek, J., Kráľová, K. (2017) *New Pesticides and Soil Sensors*. Academic Press; Cambridge, MA, USA.
48. Reddy Pullagurala, V.L., Adisa, I.O., Rawat, S., Kim, B., Barrios, A.C., Medina-Velo, I.A., Hernandez-Viezcas, J.A., Peralta-Videa, J.R., Gardea-Torresdey, J.L. (2018) Finding the conditions for the beneficial use of ZnO nanoparticles towards plants—A review. *Environmental Pollution* 241: 1175–1181.

49. Affam, A.C., Chaudhuri, M. (2013) Degradation of pesticides chlorpyrifos, cypermethrin and chlorothalonil in aqueous solution by TiO$_2$ photocatalysis. *Journal of Environmental Management* 130: 160–165.
50. Hanh, N.T., Le Minh Tri, N., Van Thuan, D., Thanh Tung, M.H., Pham, T.D., Minh, T.D., Trang, H.T., Binh, M.T., Nguyen, M.V. (2019) Monocrotophos pesticide effectively removed by novel visible light driven Cu doped ZnO photocatalyst. *Journal of Photochemistry and Photobiology A: Chemistry* 382: 111923.
51. Bourguet, D., Guillemaud, T. (2016) In: *Sustainable Agriculture Reviews.* Lichtfouse E . (Ed.), p. 120. Springer International Publishing; Cham, Switzerland.
52. Bombo, A.B., Pereira, A.E.S., Lusa, M.G., De Medeiros Oliveira, E., De Oliveira, J.L., Campos, E.V.R., De Jesus, M.B., Oliveira, H.C., Fraceto, L.F., Mayer, J.L.S. (2019) A mechanistic view of interactions of a nanoherbicide with target organism. *Journal of Agricultural and Food Chemistry* 67: 4453–4462.
53. Osorio-Echavarría, J., Osorio-Echavarría, J., Ossa-Orozco, C.P., Gómez-Vanegas, N.A. (2021) Synthesis of silver nanoparticles using white-rot fungus Anamorphous Bjerkandera sp. R1: Influence of silver nitrate concentration and fungus growth time. *Scientific Reports* 11: 1–14.
54. Vigneshwaran, N., Kathe, A.A., Varadarajan, P.V., Nachane, R.P., Balasubramanya, R.H. (2006) Biomimetics of silver nanoparticles by white rot fungus, Phaenerochaete chrysosporium. *Colloids and Surfaces: B Biointerfaces* 53: 55–59.
55. Hayles, J., Johnson, L., Worthley, C., Losic, D. (2017) *Nanopesticides: A Review of Current Research and Perspectives.* Elsevier Inc.; Amsterdam, Netherlands.
56. Khan, I., Saeed, K., Khan, I. (2019) Nanoparticles: Properties, applications and toxicities. *Arabian Journal of Chemistry* 12: 908–931.
57. Scognamiglio, V. (2013) Nanotechnology in glucose monitoring: advances and challenges in the last 10 years. *Biosensors and Bioelectronics* 47: 12–25.
58. Liu, X., Wang, R., Xia, Y., He, Y., Zhang, T. (2011) LiCl-modified mesoporous silica SBA-16 thick film resistors as humidity sensor. *Sensor Letters* 9: 698–702.
59. Zhao, C.-L., Qin, M., Huang, Q.-A. (2011) Humidity sensing properties of the sensor based on graphene oxide films with different dispersion concentrations. *IEEE Sensors Journal* 129–132.
60. Antonacci, A., Arduini, F., Moscone, D., Palleschi, G., Scognamiglio, V. (2018) Nanostructured (bio)sensors for smart agriculture. *TrAC Trends in Analytical Chemistry* 98: 95–103.
61. Mura, S., Greppi, G., Roggero, P.P., Musu, E., Pittalis, D., Carletti, A., Ghiglieri, G., Irudayaraj, J. (2015) Functionalized gold nanoparticles for the detection of nitrates in water. *International Journal of Environmental Science and Technology* 12: 1021–1028.
62. Delgadillo-Vargas, O., F-Al, Jaime, Roberto, G.-R. (2016) Fertilising techniques and nutrient balances in the agriculture industrialization transition: the case of sugarcane in the Cauca river valley (Colombia), 1943–2010. *Agriculture, Ecosystems & Environment* 218: 150–162.
63. Juárez-Cisneros, G, Gómez-Romero, M, Reyes de la Cruz, H, Campos-García, J, Villegas, J. (2020) Multi-walled carbon nanotubes produced after forest fires improve germination and development of *Eysenhardtia polystachya. PeerJ* 8: e8634.
64. Haghighi, M., Teixeira da Silva, J.A. (2014) The effect of carbon nanotubes on the seed germination and seedling growth of four vegetable species. *Journal of Crop Science and Biotechnology* 17: 201–208.
65. Ali, M.H.; Sobze, J.-M.; Pham, T.H.; Nadeem, M.; Liu, C.; Galagedara, L.; Cheema, M.; Thomas, R. (2020) Carbon nanotubes improved the germination and vigor of

plant species from peatland ecosystem via remodeling the membrane lipidome. *Nanomaterials* 10: 1852.

66. Chen, L., Guo, Z., Lao, B., Li, C., Zhu, J., Yu, R., & Liu, M. (2021). Phytotoxicity of halloysite nanotubes using wheat as a model: seed germination and growth. *Environmental Science: Nano* Issue 10.

67. Mohamed, A., Mahmoud, E., Younes, N. (2021) Impact of foliar application of carbon nanotube and benzyladenine on broccoli growth and head yield. *Archives of Agriculture Sciences Journal* 4(1): 81–101.

68. Holghoomi, R., Sarghein, S.H., Khara, J. et al. (2021) Effect of functionalized-carbon nanotube on growth indices in *Ocimum basilicum* L. grown in vitro. *Russian Journal of Plant Physiology* 68: 958–972.

69. Baz, H., Creech, M., Chen, J., Gong, H., Bradford, K., Huo, H. (2020) Water-soluble carbon nanoparticles improve seed germination and post-germination growth of lettuce under salinity stress. *Agronomy* 10: 1192.

70. Cao, Z., Zhou, H., Kong, L. et al. (2020) A novel mechanism underlying multi-walled carbon nanotube-triggered tomato lateral root formation: the involvement of nitric oxide. *Nanoscale Research Letters* 15: 49.

71. Chen, J., Zeng, X., Yang, W. et al. (2021) Seed priming with multiwall carbon nanotubes (MWCNTs) modulates seed germination and early growth of maize under cadmium (Cd) toxicity. *Journal of Soil Science and Plant Nutrition* 21: 1793–1805.

72. Liné, C., Manent, F., Wolinski, A., Flahaut, E., Larue, C. (2021) Comparative study of response of four crop species exposed to carbon nanotube contamination in soil. *Chemosphere* 274: 129854.

73. Tiwari, D.K., Dasgupta-Schubert, N., Villaseñor Cendejas, L.M. et al. (2014) Interfacing carbon nanotubes (CNT) with plants: enhancement of growth, water and ionic nutrient uptake in maize (*Zea mays*) and implications for nanoagriculture. *Applied Nanoscience* 4: 577–591.

74. Choudhary, R.C., Kumaraswamy, R.V., Kumari, S. et al. (2017) Cu-chitosan nanoparticle boost defense responses and plant growth in maize (*Zea mays* L.). *Science Reports* 7: 9754.

75. Kibbey, T.C.G., Strevett, K.A. (2019). The effect of nanoparticles on soil and rhizosphere bacteria and plant growth in lettuce seedlings. *Chemosphere* doi:10.1016/j.chemosphere.2019.01

76. Al-Amri, N., Tombuloglu, H., Slimani, Y., Akhtar, S., Barghouthi, M., Almessiere, M., Alshammari, T., Baykal, A., Sabit, H., Ercan, I., Ozcelik, S. (2020) Size effect of iron (III) oxide nanomaterials on the growth, and their uptake and translocation in common wheat (*Triticum aestivum* L.). *Ecotoxicology and Environmental Safety* 194: 110–377.

77. Venkatachalam, P., Priyanka, N., Manikandan, K., Ganeshbabu, I., Indiraarulselvi, P., Geetha, N., Muralikrishna, K., Bhattacharya, R.C., Tiwari, M., Sharma, N., Sahi, S.V. (2016) Enhanced plant growth promoting role of phycomolecules coated zinc oxide nanoparticles with P supplementation in cotton (*Gossypium hirsutum* L.). *Plant Physiology et Biochemistry* doi: 10.1016/j.plaphy.2016.09.004.

78. Younes N.A., Hassan H.S., Elkady M.F., Hamed A.M., Dawood M.F.A. (2020) Impact of synthesized metal oxide nanomaterials on seedlings production of three Solanaceae crops. *Heliyon*, 6(1): e03188.

79. Choudhary, R.C. et al. (2017). Synthesis, characterization, and application of chitosan nanomaterials loaded with zinc and copper for plant growth and protection. In: Prasad, R., Kumar, M., Kumar, V. (Eds.) *Nanotechnology*. Springer, Singapore.

80. Yoon, H., Kang, Y.-G., Chang, Y.-S., Kim, J.-H. (2019) Effects of zerovalent iron nanoparticles on photosynthesis and biochemical adaptation of soil-grown *Arabidopsis thaliana*. *Nanomaterials* 9: 1543.

81. Singh, J., Kumar, S., Alok, A., Upadhyay, S.K., Rawat, M., Tsang, D.C.W., Bolan, N., Kim, K.-H. (2019) The potential of green synthesized zinc oxide nanoparticles as nutrient source for plant growth. *Journal of Cleaner Production* doi:10.1016/j.jclepro.2019.01.018.

82. Hu, J., Guo, H., Li, J. et al. (2017) Interaction of γ-Fe_2O_3 nanoparticles with *Citrus maxima* leaves and the corresponding physiological effects via foliar application. *Journal of Nanobiotechnology* 15: 51.

83. Kumar, R., Ashfaq, M., Verma, N. (2018) Synthesis of novel PVA–starch formulation-supported Cu–Zn nanoparticle carrying carbon nanofibers as a nanofertilizer: controlled release of micronutrients. *Journal of Materials Science* 53: 7150–7164.

84. Jiang, F., Shen, Y., Ma, C., Zhang, X., Cao, W., Rui, Y. (2017) Effects of TiO_2 nanoparticles on wheat (*Triticum aestivum L.*) seedlings cultivated under super-elevated and normal CO_2 conditions. *PLoS ONE* 12(5): e0178088.

85. Shekhawat, G.S., Mahawar, L., Rajput, P., Rajput, V.D., Minkina, T., Singh, R.K. (2021) Role of engineered carbon nanoparticles (CNPs) in promoting growth and metabolism of *Vigna radiata* (L.) Wilczek: Insights into the biochemical and physiological responses. *Plants* 10: 1317.

86. Iqbal, M.A. (2019) Nano-fertilizers for sustainable crop production under changing climate: A global perspective. In: M. Hasanuzzaman, M.C.M.T. Filho, M. Fujita, T. A. R. Nogueira (Eds.) *Sustainable Crop Production*. IntechOpen.

87. Das, K., Jhan, P.K., Das, S.C., Aminuzzaman, F., Ayim, B.Y. (2021) Nanotechnology: Past, present and future prospects in crop protection. In: Ahmad, F., Sultan, M. (Eds.) *Technology in Agriculture*. London: IntechOpen. Available from: www.intechopen.com/chapters/77333.

88. Xu, L., Liu, Y., Bai, R., Chen, C. (2010) Applications and toxicological issues surrounding nanotechnology in the food industry. *Pure and Applied Chemistry* 82: 349–372.

89. Chen, H., Yada, R. (2011) Nanotechnologies in agriculture: New tools for sustainable development. *Trends in Food Science & Technology* 22: 585–594.

90. Han, C., Zhao, A., Varughese, E., Sahle-Demessie, E.J.N. (2018) Evaluating weathering of food packaging polyethylene-nano-clay composites: Release of nanoparticles and their impacts. *NanoImpact* 9: 61–71.

91. Narei, H., Ghasempour, R., Akhavan, O. (2018) Toxicity and safety issues of carbon nanotubes. In: *Carbon Nanotube-Reinforced Polymers: From Nanoscale to Macroscale*; Rafiee, R. (Ed.), pp. 145–171. Elsevier: Amsterdam, The Netherlands.

92. Pathakoti, K., Manubolu, M., Hwang, H.-M. (2017) Nanostructures: Current uses and future applications in food science. *Journal of Food and Drug Analysis* 25: 245–253.

93. Kumar, S., Shukla, A., Baul, P.P., Mitra, A., Halder, D. (2018) Biodegradable hybrid nanocomposites of chitosan/gelatin and silver nanoparticles for active food packaging applications. *Food Packaging Shelf Life* 16: 178–184.

94. Li, L., Zhao, C., Zhang, Y., Yao, J., Yang, W., Hu, Q., Wang, C., Cao, C. (217) Effect of stable antimicrobial nano-silver packaging on inhibiting mildew and in storage of rice. *Food Chemistry* 215: 477–482.

95. Sharma, C., Dhiman, R., Rokana, N., Panwar, H. (2017) Nanotechnology: An untapped resource for food packaging. *Frontiers in Microbiology* 8: 1735.

96. Oleszczuk, B.-W.E.S.D. Effects of titanium dioxide nanoparticles exposure on human health—A review. *Biological Trace Element Research* 193: 118–129.

97. McClements, D.J. (2020) Advances in nanoparticle and microparticle delivery systems for increasing the dispersibility, stability, and bioactivity of phytochemicals. *Biotechnology Advances* 38: 107287.

98. Kumar, S., Nehra, M., Dilbaghi, N., Marrazza, G., Tuteja, S.K., Kim, K.H. (2020) Nanovehicles for plant modifications towards pest- and disease-resistance traits. *Trends in Plant Science* 25: 198–212.

99. Demirer, G., Zhang, H., Goh, N., Chang, R., Landry, M. (2019) Nanotubes effectively deliver siRNA to intact plant cells and protect siRNA against nuclease degradation. *Biology Engineering.*

5 Microbial Enzymes

Current Features and Potential Applications in Bionanotechnology

Juhi Sharma, Divakar Sharma, Karan Sharma, Surabhi Sharma, Priya Choudhary, and Akshay Bharti

5.1 INTRODUCTION

The creation, characteristics, and shape of materials at the nanoscale are the focus of nanotechnology. The emergence of various characteristics in materials become more significant as they approach the nanoscale. Applications for nanomaterials may be found in a wide range of sectors, including energy storage, space sciences, biomedicine, catalysis, etc. Due to their shape- and size-dependent characteristics, nanoparticles have garnered a great deal of interest. The synthesis of fluorescent nanoparticles using enzymes is one such recent application. Mansoori et al., 2007 studied the in vitro synthesis of fluorescent nanoparticles using enzyme reductases.

5.2 APPLICATION OF ENZYME NANOPARTICLES IN LIFE AND APPLIED SCIENCES

It is already known that all biological bodies, such as living cells and various microorganisms, are examples of machines which have active parts at the micro-scale level which carry out various tasks, starting from energy formation to the extraction of targeted materials, with a very high level of efficiency. It is a relatively new practice to include microorganisms, such as bacteria, fungi, yeasts, and herbal concentrates, in the creation of nanoparticles. As long as they are not dangerous in other ways, certain bacteria, yeasts, and now fungi can play a significant part in refining toxic metals by reducing the presence of metal ions (Goodsell, 2004).

By reducing the metal ions or creating an insoluble complex with metal ions (such as sulfide metal) we can formulate colloidal particles, for instance, environment-friendly microorganisms could reduce the toxicity by producing metallic nanoparticles. Therefore, eco-friendly biological bodies may be regarded as safe nanofactories. It should be noted that various microbes among these are physiologically toxic for humans, plants, and animals, thus caution should be exercised when selecting them for the manufacture of nanoparticles. The necessary aspect of

DOI: 10.1201/9781003362258-5

the constantly expanding research activities in nano-scale science and engineering is nanoparticle manufacturing. The synthesis of nanoparticles using biotechnology has various benefits, including the simplicity with which the process can be scaled up, its practicality from an economic standpoint, and its potential for readily covering large areas (Mehra and Winge, 1991).

Starch degradation or hydrolysis by amylase shows a stimulant behavior of AgNPs in an enzyme-catalyzed process. By immobilizing the enzymes on the surface of AgNPs, this study offers a route to the speedier breakdown of starch, which could be a viable use in the food industry. Doping this hybrid with silver nanoparticles (AgNPs) might greatly lengthen the enzyme's shelf life while maintaining its complete biocatalytic activity. Using a unique photosensitive microemulsion polymerization technique, nanoprotein particles carrying α-amylase as a monomer have been described. The primary impacts on various seeds were assessed in regard to biotic and abiotic stresses, as well as the physico-chemical characteristics of the nanoparticles employed for seed priming (Ahmad and Sardar, 2014).

It is possible to boost a crop's resistance to stress at the early seedling stage with PNC nanoparticle seed priming in a sustainable, useful, and scalable manner. These findings suggested interesting biochemical processes that may work in concert to increase a plant's resistance to salt. Under salt stress, nano-priming using water-dissolving carbon nanoparticles (CNPs) greatly boosts lettuce seed vigor and seedling development. It has been demonstrated that CNPs greatly increase seed germination when exposed to high temperatures and saline stress. A balanced buildup of chlorophyll under severe salinity stress was achieved by a CNP-assisted nanopriming therapy, which improved lateral root development but only marginally hindered elongation of the main roots (Figure 5.1).

By nano-priming with a range of nanomaterials, fertilization of seeds and flowering growth parameters of numerous species of plants are improved (Misson et al., 2015). A unique strategy for sustainable agriculture has been found in a number of research works on the interconnection between nanomaterials and bacteria that benefit plants, known as plant growth-promoting rhizobacteria (PGPR). Plants can survive environmental stress by a variety of morphological and physiological processes,

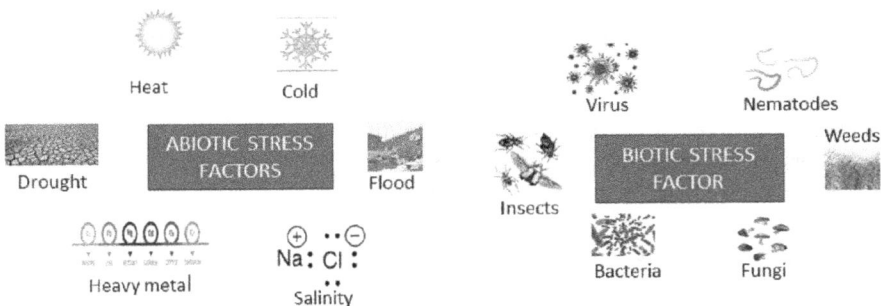

FIGURE 5.1 Nanocatalyst to reduce abiotic and biotic stress for sustainable agriculture development.

including better root system nutrient absorption and elevated production of cellular antioxidative enzymes, when both PGPR and nanomaterials are applied. Mahakham et al. investigated the work of silicon nanoparticles in the cell development procedure employing silicon nanoparticles as a seed priming agent. This investigation looked at how seed priming with silicon nanoparticles affected the growth, physiology, and antioxidant capacities of *Pseudomonas* species (Kuan et al., 2018).

A *P. agglomerans* seed covering increased root weight, growth, and germination. Additionally, seed coating with *Bacillus californianus* boosted the dry weight and number of leaves. Due to this, the method used in current work to key seeds with PGPR immobilized on nanofibers may be considered a hopeful environment-friendly method to increase the yield of soybean by employing microbial infusion (Misson et al., 2015). Taran et al. claimed that a solution of metals for micronutrients improved plant health by boosting resilience in its vulnerability to harsh environment circumstances, and also improved the nutrition of plants by allowing an increase in penetration of nanoscale elements in cell walls. When seed treatment is combined with a colloidal solution of Mo nanoparticles with a microbial preparation, nodule development was increased four times more than in control plants.

Compared to the healthcare and industrial areas, utilizing nanomaterials in agriculture is relatively new. In order to further environmentally friendly nanoagriculture, nanocomposites of silver nanoparticles (AgNPs) have been created using kaffir lime leaf extract. These AgNPs are used as nano priming agents to improve the vegetation of rice-aged seeds. The effective production of AgNPs crowned with phytochemicals from the plant extract was demonstrated by the results of several characterization procedures (Gupta et al., 2018). Researchers' interest in using environmentally friendly methods to create metal nanoparticles has increased as a result of the widespread usage of these particles in home goods and medical equipment. Plants may be used to create nanoparticles, which is both economical and environmentally benign. AgNPs were produced through a straightforward and biosynthetic technique (Mohammadi et al., 2022).

The diameters of the AgNPs produced during biosynthesis varied with reaction time. As a result, AgNP size rises at room temperature as the reaction time increases. Elastic plays a notable role in the bioreduction and stabilization of the silver ions as AgNPs, as shown by the findings of this work. An innovative new product called nanobiocatalyst (NBC) provides promising benefits for enhancing the capability, stability, enzyme-activity, and engineering performances in the application of bioprocessing. It merges sophisticated nanotechnology and biotechnology synergistically. In order to immobilize enzymes, functional nanomaterials are used as containers for the enzymes (Mohammadi et al., 2022).

The most promising biomaterials created by combining modern nanotechnology and biotechnology are called nanobiocatalysts. These offer a great deal of promise to enhance the stability, performance, and engineering capabilities of enzymes in bioprocessing. Their unique biophysical properties and macromolecular nature, and working nanostructures are exploited in the making of nanobiocatalysts (Mohammadi et al., 2022).

As a result, their use as natural catalysts is limited. Nanobiocatalysts, produced through the combination of nanotechnology and biotechnology, are a result of using nanotechnology and also the potential for enzyme immobilization on nanomaterials. To increase the activity, stability, efficiency, and storage stability of enzymes, nanocarriers have unique properties such as microscale size, outstanding surface and volume ratios, and a variety of designs (Mukherjee et al., 2003). There are many applications for nanobiocatalysts in manufacturing, packaging, clarifying, extraction, and purification (Verma et al., 2016) (Figure 5.2).

Chemists, engineers, and material scientists can collaborate to find the solutions to problems. Finally, by fruitful lab tests of nanobiocatalysts in carbohydrate hydrolysis, biofuel generation, and biotransformation, enormous potential has been shown of NBCs in producing bioprocesses for the future.

A multitude of other nanoscale carriers have recently become available thanks to advances in nanotechnology, which could be used for enzyme immobilization. A potential strategy to improve enzyme performances has been identified as enzyme immobilization on nanostructured materials (Reshmy et al., 2021). A new invention that successfully combines advances in nanotechnology and biotechnology is called nanobiocatalyst (NBC). NBC construction includes the assembly of enzyme molecules onto carriers made of nanomaterials in sequence to promote the desired chemical kinetics and substrate selectivity. Enzymes may be put together in sequenced structures that serve as an information storage and processing system in the nanoscale thanks to the functionalized nanocarriers. In research, it has been shown that nanostructured materials acquire the properties of NBCs and support huge surface

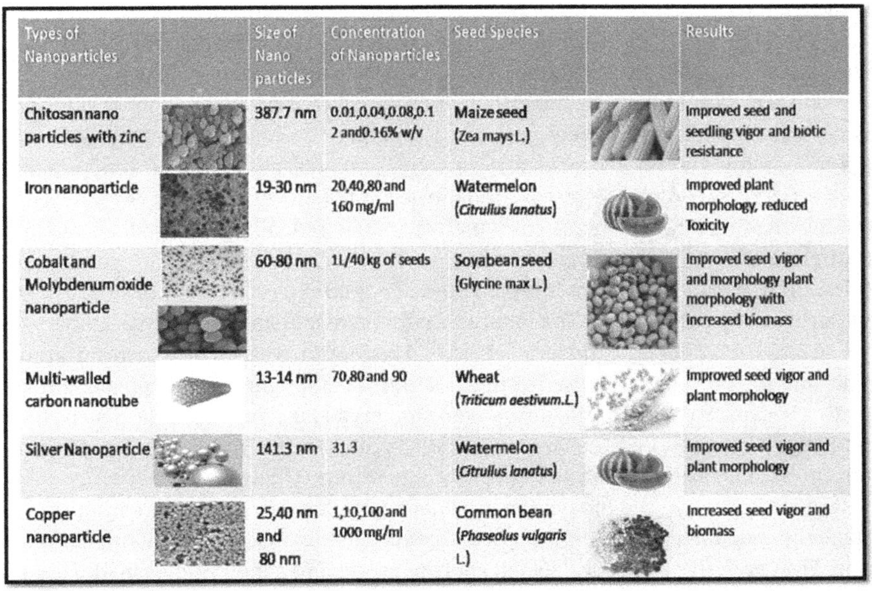

Types of Nanoparticles		Size of Nano particles	Concentration of Nanoparticles	Seed Species		Results
Chitosan nano particles with zinc		387.7 nm	0.01,0.04,0.08,0.1 2 and0.16% w/v	Maize seed (Zea mays L.)		Improved seed and seedling vigor and biotic resistance
Iron nanoparticle		19-30 nm	20,40,80 and 160 mg/ml	Watermelon (Citrullus lanatus)		Improved plant morphology, reduced Toxicity
Cobalt and Molybdenum oxide nanoparticle		60-80 nm	1l/40 kg of seeds	Soyabean seed (Glycine max L.)		Improved seed vigor and morphology plant morphology with increased biomass
Multi-walled carbon nanotube		13-14 nm	70,80 and 90	Wheat (Triticum aestivum.L)		Improved seed vigor and plant morphology
Silver Nanoparticle		141.3 nm	31.3	Watermelon (Citrullus lanatus)		Improved seed vigor and plant morphology
Copper nanoparticle		25,40 nm and 80 nm	1,10,100 and 1000 mg/ml	Common bean (Phaseolus vulgaris L.)		Increased seed vigor and biomass

FIGURE 5.2 Applications of bionanocatalysts.

areas which enables high enzyme loading and decreases the mass transfer resistance for substrates (Verma et al., 2013).

Because they use less chemicals and do not produce any harmful byproducts, biocatalysts enable the use of green processes. However, when they are used in large-scale industries, the enzymes' poor stability and reusability result in significant operational costs for the enzyme-catalyzed bioprocesses. Using immobilized enzymes in bioreactors, however, may result in reduced activity of enzymes because immobilization alters the distinct and native structures of the enzymes (Reshmy et al., 2021). Nevertheless, we can maintain the stability and enzyme activity through precise immobilization with research for determining the best immobilization protocols.

It is thought that immobilization on solid supports may decrease the activity of some enzymes, including penicillin acylase, β-galactosidase, bovine serum albumin, and penicillin acylase. Depending on how enzymes are mounted on nanocarriers, this notion may not be accurate. Numerous research works have been carried out on how adding nanocarriers to an enzyme might increase its activity.

The higher and sustained enzyme activity can be attributed to the intensified stability of nanocarrier–enzyme complexes. As a result of hydrophobic surface interactions, enzymes can become denatured. The creation of collections at high temperatures and isoelectric points, or in the presence of solvents or salts, can also cause enzymes to lose their activation (Reshmy et al., 2021).

Furthermore, due to their excellent detection accuracy, nanobiocatalysts have been successfully used to pinpoint particular elements of food pollutants such as metabolites of bacteria, heavy metals, residual pesticides, and antibiotics (Mohammadi et al., 2022). The incorporation of nanotechnology and food enzymes, the creation and usage of nanomaterials used to make nanobiocatalysts, pose potential dangers and ethical considerations for using nanomaterials in food bioprocesses (Bilal et al., 2021).

As a brand-new field of study, biofuels are quickly progressing to offer substitute clean and sustainable energy sources. In the latest developments in nanotechnology, it has been aimed at increasing the effective production of biofuels, thereby improving energy security. Enzymes may be placed together in ordered structures that serve as a processing system and nanoscale information storage, thanks to functionalized nanocarriers. According to research, nanostructured materials meet the criteria for NBCs by having huge surface areas which enable more loading of enzymes and lower substrate mass transfer resistance.

In-depth research into bacteria-mediated nanoparticle production has been carried out during the past 10 years. The first report on the manufacture of silver nanocrystals was using *Pseudomonas stutzeri* A25961. This strain was identified from a silver mine, and the presence of plasmids in the strain was thought to be responsible for the bacteria's capacity to manufacture silver nanocrystals. Even bacteria that have not previously been exposed to hazardous metals, such as *Lactobacillus* sp. present in buttermilk, have been known to crystallize silver ions (Chawla et al., 2013).

To a large degree, the biosynthesis of nanoparticles utilizing microorganisms has satisfied the demand for an environmentally acceptable method for nanomaterial creation. However, these approaches fall short of producing nanoparticles in ideal shapes and sizes, which is critical for realizing the full potential of these nanomaterials.

To better understand the principles underlying nanomaterial production, researchers have used biomolecules to create nanomaterials, simplifying the process. The possibility of adjusting the technique for a better product is increased by determining the role of each component in the process (Ovais et al., 2018).

Researchers have searched the sources of nanoparticles for biomolecules that are responsible for nanoparticle synthesis. Except for some plant extract-based synthesis techniques, they have discovered that the biomolecules which are responsible for nanoparticle formation are predominantly enzymes. Enzymes produce nanoparticles principally by reducing the matching metal ions in solution. Capping molecules then bind the reduced metal ions, avoiding agglomeration (Ajitha et al., 2015).

The regulated production of AgNPs for using *F. oxysporum* cell-free filtrate has been investigated. When the parameters of nitrogen and carbon sources, presence or absence of light, and cultivation temperature during cell growth on formation of AgNP were considered, the outcomes revealed that there was an increase at the stationary phase for nanoparticle formation, which is due to the increase in secretion of extracellular enzymes, particularly the NR enzyme (Wang et al., 2010). Controlling the growth of fungi at 28°C may result in creating acceptable AgNP sizes. The latency of glucose in the MGYP medium, the supply of nitrogen in changed medium, and light all produce comparable outcomes. In the case of a fungal culture in modified medium with maximal activity of the NR enzyme, the smallest AgNPs with the best productivity were obtained (Ajitha et al., 2015).

Due to their intriguing biological, physical, magnetic, and optical features, metal-based nanoparticles have garnered enormous interest. These nanoparticles may be created utilizing a range of biological, physical, and chemical methods. Among all the methods, the biological method is widely favored since it is ecologically friendly, cost-effective, and green for producing the nanoparticles. The function of these bioresources is as scaffolds, acting as both capping and reducing agents in the biogenesis of nanoparticles. Medicinal plants include complex phytochemical constituents such as alcohols, terpenes, alkaloids, saponins, phenols, and proteins, whereas microorganisms have essential enzymes that may function as reducing and stabilizing agents for NP synthesis (Ovais et al., 2018).

The mechanism of biosynthesis, on the other hand, is still extensively debated. The purpose of this study is to update the complete account of mechanistic elements of nanoparticle production using plants and microorganisms. Various secondary metabolite biosynthesis routes into plants and important production of enzymes in microorganisms have been rigorously explored, as have the underlying processes for biogenic NP synthesis (Nangia et al., 2009).

Production of the gold nanoparticles (AuNPs) has received a lot of attention because of their prospective uses in several life sciences. There has recently been a lot of interest in research into nanoparticle synthesis utilizing natural biological systems, which has led to the development of several biomimetic techniques for the generation of sophisticated nanomaterials (Ahmad et al., 2013). We have illustrated showed gold nanoparticle manufacture using a unique strain of bacteria obtained from a location close to India's famous gold mines in this work. The possible method for this strain's biosynthesis of AuNPs and their stability via charge capping were examined (Ovais et al., 2018).

5.3 APPLICATION OF ENZYME NANOPARTICLES IN FORENSIC SCIENCE

Forensic scientists are concerned with the evidence found at the scene of a crime and interpretation of the results from the available evidence is based solely on the standard operating procedures followed in the examination. Most of the time, the evidence at the scene of a crime is in very small quantities or trace amounts, and analyzing such small amounts of evidence can become a difficult task for the examiner. To resolve such problems, nanotechnology has been proved to be an excellent option. In the past decade, many research works has been carried out by various researchers which have exhibited suitable application of nanotechnology in this investigative discipline of forensic science (Brayner et al., 2007). The applications of nanotechnology in forensic science includes the detection of warfare agents, spotting drugs in alleged crimes, gunshot residues, explosives and post-blast residue detection, latent fingerprint development, DNA analysis, etc. (Chakraborty et al., 2015).

The detailed uses of these nanoparticles in forensic science are described below.

5.3.1 DETECTION OF WARFARE AGENTS

Warfare agents are harmful chemical and biological materials which can be fatal for humans, and they emerged as a mass destruction weapon during the world wars and have been used in a great extent in conflicts. Having a noxious characteristic, these agents can be biological, chemical, or radiological, and spread rapidly in the environment causing a large number of causalities. The action of warfare agents on humans can be fatal and threaten the basic functions of life such as sight, respiration, circulation, etc.

Thus, these poisonous warfare agents need to be detected before they enter the atmosphere and create a threat to human life, which is where nanoparticles come into use as they help in detecting these agents by immediate sensor-based colorimetric and fluorometric analysis, while also helping in making the toxins non-reactive (Chakraborty et al., 2015).

Nanoparticles are used in making sensors for the detection, destruction, tracing, and decontamination of chemical, biological, and radiological warfare agents. As per the nature of the warfare agents, nanoparticles are implanted in microchips and these microchips help in the detection of the specific agent by changing its color as it comes into contact with the specific ligand molecule. It can even prevent the effects of toxic agents by covering the toxin and making a non-toxic byproduct, thus making the environment non-toxic for humans (García-Briones et al., 2019).

The most commonly used nanoparticles in detecting such agents are gold (Au) ones. Because there is the presence of free electrons on the metal surface, AuNPs have a high surface plasmon resonance property and, on light exposure, these free electron absorbs the electromagnetic spectrum via the visible and near-infrared region. On assemblage of AuNPs, surface plasmon coupling occurs which shifts from the visible region to the IR region, thus helping in the phenomenon of changing color. AuNPs in combination with different sensors and compounds help in detecting various warfare agents, some of which are mentioned below.

1. The toxic biological agent ricin is detected using AuNPs with an aptasensor qualitative analytical procedure that helps by showing a change in color from light blue to dark blue (Wu et al., 2023).
2. Phosgene gas is also detected using AuNPs modified with cysteine.
3. AuNPs probed with lanthanum (La^{3+}) are used to detect methyl parathion.
4. AuNPs capped with ascorbic acid are used to detect dichlorvos.
5. Toxic agents like sarin (GB), paraoxones, Vx (venomous agent X), soman (GD), etc. are also detected using AuNPs in combination with acetylcholine (Husain, 2017).

5.3.2 DEVELOPMENT OF LATENT FINGERPRINTS

Fingerprints are considered as unique characteristics of individuals which helps in the identification of culprits from a crime scene. Fingerprints are unique for everyone and start forming at the fetal stage during the 10th week and are fully developed at the fourth month of the fetal stage and then remain unchanged throughout the lifetime. As per Locard's principle of exchange, the prints are transferred to the surface when a person comes in contact with it, which makes it easy for an investigator to catch the culprit using their unique fingerprints. These prints can get diminished or destroyed over time if the prints are not lifted properly and quickly (Wu et al., 2023).

Usually a powder method (white, black, magnetic, luminescent powder) is used to lift the fingerprints for further analysis but, since the particle size of these powders is large, apart from the fingerprints, the background also gets stained, which makes it very difficult for a forensic scientist to examine it, and that is where nano particles are used in the analysis as their particle size is small, which increases the surface area of absorption for the fingerprints. Some of the nanoparticles used in latent fingerprint development are mentioned below.

1. Titanium oxide (TiO) nanoparticles have been recently used nano particles in the lifting of prints from porous as well as non-porous surfaces (Prasad et al., 2019).
2. Zinc oxide nanoparticles (ZnO) are also used to lift prints in wet climatic conditions.
3. Petroleum ether suspended AuNPs stick tightly to latent fingerprints and after immersion in silver ion solution give a clear silver outline to visualize the latent fingerprints (Martí et al., 2014).
4. To know if the person is a smoker or not from its fingerprints AuNPs combined with cotinine-specific antibodies (nicotine metabolite) are used.

5.3.3 ANALYSIS OF DRUGS

Drugs are commonly used in many serious crimes such as rape, sexual assault, murder, extortion, burglary, etc. These illegal drugs needed to be identified as soon as possible to aid law enforcement agencies. With the use of conventional methods, rapid identification of drugs is not possible. As these methods require proper laboratory setup, are quite costly, time consuming, and with the unavailability of instruments, the analysis can be delayed, which will hamper the investigation process.

Rapid and on-site drug analysis is carried out using nanotechnology. Devices like nanoprobes, nanochips, and nanosensors are cost-effective and have high sensitivity in the detection of trace amounts of these illegal drugs. Some of the common illicit drugs that are used in alleged drug-facilitated crimes that require rapid detection include ketamine, amphetamines, opioids, benzodiazepines, MDMA, flunitrazepam, GHB, rohypnol, and cocaine.

Nanoparticles used in drug analysis are:

1. AuNPs accompanied with citrate help in the identification of drugs like codeine.
2. Clonazepam can be identified by AuNPs mixed with melamine (Choi et al., 2008).
3. Gold nanoparticles and silver nanoparticles in combination with the FTIR method can be used to detect fingerprints with cocaine (Lad et al., 2016).

5.3.4 ANALYSIS OF EXPLOSIVES AND POST-BLAST RESIDUE

Explosions are a devastating method of causing death and destruction used by terrorists. They have been proven to be one of the most suitable methods for mass destruction. These explosives can be detected if found unexploded hidden in containers such as soft-toy, pressure cookers, tiffin, suitcases, etc., and also detection of explosives can be done from the post-blast residue.

Nanotechnology-based devices such as nanosensors and nanotubes, and electronic and mechanical nanodevices such as the electronic nose are used to detect such explosive materials (Srividya, 2016). Silver and gold nanoparticles in combination with SERS (surface-enhanced Raman scattering spectroscopy) are used in the detection of many explosive materials like TNT, PETN, RDX, etc. (Venditti et al., 2015).

5.3.5 DNA ANALYSIS

The discovery of DNA by scientists including Friedrich Miescher, James Watson, and Fransic Crick, and the work carried out by Sir Alec Jefferys opened a vast field for individualized identification of humankind. DNA analysis is one of the most precise methods for the identification of offenders in crimes like sexual assault, rape, murder, etc. The presence of DNA in almost all biological samples is of utmost importance when it comes to the identification of individuals using biological samples. DNA analysis involves the study of genetic markers such as SNP, STR, and VNTR sequences with the aim of the generation of the complete genetic profile of the individual to identify them (Pandya and Shukla, 2018).

For identification purposes, the DNA needs to be isolated from the biological samples that are collected from the scene of the crime, and it is very difficult task to extract DNA from such small amounts of samples of saliva, semen, blood, etc. Magnetic nanoparticles such as Fe_3O_4 when used as a solid medium can be utilized for fast and effective extraction of DNA from small quantities of biological samples including hair, saliva, semen, etc. (Saiyed et al., 2008).

5.4   CONCLUSION

The emerging science of nanotechnology has illustrated an important role in examining trace evidence, both volatile and non-volatile, and hence helping in forensic investigations where other techniques have been inconclusive. In-depth study is required to enable the huge potential in bioprocessing of nanobiocatalysts for creating efficient lab tests for applications in a variety of industries, including food, medicines, biofuel, and bioremediation, and also in forensic science. New green technologies that are superior to chemical procedures in the genetic analysis sector have been inspired by the rapid advancements in current technology and the increased focus on forensic detection. Enzymes and nanoparticles can reduce the need for chemical reactions in this way, but are susceptible to various factors in the environment (temperature and pH), which still need to be explored for sustainable development.

REFERENCES

Ahmad, R., Sardar, M. (2014) Immobilization of cellulase on TiO_2 nanoparticles by physical and covalent methods: a comparative study. *Indian Journal of Biochemistry Biophysics* 51(4): 314–320.

Ahmad, R., Mishra, A., Sardar, M. (2013) Peroxidase-TiO_2 nanobioconjugates for the removal of phenols and dyes from aqueous solutions. *Advance Science Engineering Medicine* 5: 1020–1025.

Ajitha, B., Ashok, Kumar, Reddy, Y., Shameer, S., Rajesh, K.M., Suneetha, Y., Sreedhara, Reddy, P. (2015) Lantana camara leaf extract mediated silver nanoparticles: antibacterial, green catalyst. *Journal of Photochemistry and Photobiology B: Biology* 149: 84–92.

Bilal, M., Hussain, N., Américo-Pinheiro, J.H.P., Almulaiky, Y.Q., Iqbal, H.M.N. (2021) Multi-enzyme co-immobilized nano-assemblies: Bringing enzymes together for expanding bio-catalysis scope to meet biotechnological challenges. *International Journal of Biological Macromolecules* 186: 735–749.

Brayner, R., Barberousse, H., Hemadi, M., Djediat, S., Yepremian, C., Coradin, T., Livage, F.F., Coute, A. (2007) Cyanobacteria as bioreactors for the synthesis of Au, Ag, Pd, and Pt nanoparticles via an enzyme-mediated route. *Journal of Nanoscience Nanotechnology* 7: 2696–2708.

Chakraborty, D., Rajan, G., Isaac, R. (2015) A splendid blend of nanotechnology and forensic science. *Journal of Nanotechnology in Engineering and Medicine* 6.

Chawla, S., Rawal, R., Sonia, R., Pundir, C.S. (2013) Preparation of cholesterol oxidase nanoparticles and their application in amperometric determination of cholesterol. *Journal of Nanoparticle Research* 15: 19341943.

Choi, M., McDonagh, A., Maynard, P., Roux, C. (2008) Metal-containing nanoparticles and nano-structured particles in fingermark detection. *Forensic Science International* 179(2–3): 87–97.

Deccan Chronicle (2019) *Nanotechnology on the crime scene..* www.deccanchronicle.com/nation/current-affairs/010419/nanotechnology-on-the-crime-scene.html.

García-Briones, G., Olvera-Sosa, M., Palestino, G. (2019) Novel supported nanostructured sensors for chemical warfare agents (CWAs) detection. *Nanoscale Materials For Warfare Agent Detection: Nanoscience For Security* 225–251.

Goodsell, D.S. (2004). *Bionanotechnology: Lessons from Nature.* Wiley-Liss, Hoboken, New York.

Gupta, M., Tomar, R. S., Kaushik, S., Mishra, R. K., Sharma, D. (2018) Effective antimicrobial activity of green ZnO nano particles of *Catharanthus roseus. Frontiers in Microbiology* 9: 2030.

Husain, Q. (2017) Nanomaterials as novel supports for the immobilization of amylolytic enzymes and their applications: a review. *Biocatalysis* 3: 37–53.

Lad A, Pandya A, Agrawal Y. (2016) Overview of nano-enabled screening of drug-facilitated crime: A promising tool in forensic investigation. *TraC Trends in Analytical Chemistry* 80: 458–470.

Mansoori, G.A., George, T.F., Zhang, G., Assoufid, L. (2007) *Molecular Building Blocks for Nanotechnology.* Springer, New York.

Martí, A., Costero, A.M., Gaviña, P., Parra, M. (2014) Triarylcarbinol functionalized gold nanoparticles for the colorimetric detection of nerve agent simulants. *Tetrahedron Letters* 55(19): 3093–3096.

McDonald G. (2015) *4th International Conference on Forensic Research & Technology.* http://dx.doi.org/10.4172/2157-7145.C1.018.

Mehra, R.K., Winge, D.R. (1991) Metal ion resistance in fungi: Molecular mechanisms and their regulated expression. *Journal of Cell Biochemistry* 45: 30–40.

Misson, M., Zhang, H., Jin, B. (2015) Nanobiocatalyst advancements and bioprocessing applications. *Journal of the Royal Society Interface* 6;12(102): 20140891.

Mohammadi, Z.B., Zhang, F., Kharazmi, M.S., Jafari, S.M. (2022) Nano-biocatalysts for food applications; immobilized enzymes within different nanostructures. *Critical Reviews in Food Science and Nutrition* 25: 1–19.

Mukherjee, P., Senapati, S., Mandal, D., Ahmad, A., Islam, Khan, M., Kumar, R. (2003) Extracellular synthesis of gold nanoparticles by the fungus *Fusarium oxysoporum. ChemBioChem* 5: 461–463.

Nangia, Y., Wangoo, N., Goyal, N. (2009) A novel bacterial isolate *Stenotrophomonas maltophilia* as living factory for synthesis of gold nanoparticles. *Microbial Cell Factories* 8: 39.

Ovais, M., Khalil, A.T., Islam, N.U. et al. (2018) Role of plant phytochemicals and microbial enzymes in biosynthesis of metallic nanoparticles. *Applied Microbiology and Biotechnology* 102: 6799–6814.

Pandya, A., Shukla, R. (2018) New perspective of nanotechnology: role in preventive forensic. *Egyptian Journal Of Forensic Sciences* 8(1).

Prasad, V., Lukose, S., Agarwal, P., Prasad, L. (2019) Role of nanomaterials for forensic investigation and latent fingerprinting—A review. *Journal of Forensic Sciences* 65(1): 26–36.

Reshmy, R., Philip, E., Sirohi, R., Tarafdar, A., Arun, K,B., Madhavan, A., Binod, P., Kumar, Awasthi, M., Varjani, S., Szakacs, G., Sindhu, R. (2021) Nanobiocatalysts: Advancements and applications in enzyme technology. *Bioresource Technology* 337: 125491.

Saiyed, Z., Ramchand, C., Telang, S. (2008) Isolation of genomic DNA using magnetic nanoparticles as a solid-phase support. *Journal of Physics: Condensed Matter* 20(20): 204153.

Srividya, B. (2016).. *Nanotechnology in Forensics and its Application in Forensic Investigation.* Department of Pharmaceutical Analysis, Yalamarthy Pharmacy College, Andhra University, Vishakapatnam, India

Venditti, I., Palocci, C., Chronopoulou, L., Fratoddi, I., Fontana, L., Diociaiuti, M., Russo, M.V. (2015) *Candida rugosa* lipase immobilization on hydrophilic charged gold nanoparticles as promising biocatalysts: activity and stability investigations. *Colloids Surface B: Biointerfaces* 131: 93–101.

Verma, M.L., Chaudhary, R., Tsuzuki, T., Barrow, C.J., Puri, M. (2013) Immobilization of ß-glucosidase on a magnetic nanoparticle improves thermostability: application in cellobiose hydrolysis. *Bioresource Technology* 135: 2–6.

Verma, M.L., Puri, M., Barrow, C.J. (2016) Recent trends in nanomaterials immobilised enzymes for biofuel production. *Critical Reviews in Biotechnology* 36(1): 108–119.

Wang, H., Li, J., Wang, Y., Jin, J., Yang, R., Wang, K., Tan, W. (2010) Combination of DNA ligase reaction and gold nanoparticle-quenched fluorescent oligonucleotides: a simple and efficient approach for fluorescent assaying of single-nucleotide polymorphisms. *Analytical Chemistry* 82(18): 7684–7690.

Wu, Y., Feng, J., Hu, G., Zhang, E., Yu, H.-H. (2023) Colorimetric Sensors for Chemical and Biological Sensing Applications. Sensors 2023, 23, 2749. https://doi.org/10.3390/s23052749

6 Bionanotechnology in the Environment

Shraddha Dwivedi, Anupriya Misra,
Shreya Tiwari, Princi Tiwari, Diksha Upreti,
and Anuj Tewari

6.1 INTRODUCTION

Both domestic animals and humans share the same ecosystem. Hence, it is paramount that a healthy environment be present to promote safe human and animal interactions. In such a scenario, maintaining a sterile environment at several checkpoints is important to ensure the good health of both animals and animal care workers. Over the past few decades, nanotechnology has emerged as an important domain in diagnostics. It is becoming a new focus in the fields of medical sciences and bioengineering to contribute toward the advancement of disease diagnosis. Since the efficacy of every diagnostic method is based on factors such as sensitivity, selectivity, and economic feasibility, nanotechnology helps as it can alter the materials at the nanoscale and hence improve their ability to diagnose. Some examples are the use of gold nanoparticles, protein chips, gene chips, nanowires, and nanotubes to correctly detect the causative biomolecules of a given disease (Savaliya et al., 2015). Gold nanoparticles are the most flexible nanoparticles that can even change color on aggregation (Mekonnen, 2021). In the modern era, nanotechnology plays a crucial role as it offers diversified roles in the fields of therapeutics, antimicrobials, antibiofilm agents, drug/vaccine-delivery systems, food safety, air and water cleansing, waste management, and disinfection of veterinary care instruments/implants. Hence, the role of nanobiotechnology in providing a sterile and safe environment for the animal and human environment is discussed in detail in this chapter.

6.2 NANOTECHNOLOGY AND ANIMAL HEALTH AND PRODUCTION

6.2.1 Disease Diagnosis

Bionanomaterial is an essential aspect of nanotechnology wherein DNA, RNA, and protein entities are accepted as important bionanomaterials. All of these immensely help in the early and precise diagnosis of animal diseases. The shortcomings of the various assays such as polymerase chain reaction and western blot assay are overcome with the use of nanochips (Jo et al., 2010). Nanosensors, which are of two types, catalytic and affinity sensors, make a major contribution to veterinary sciences

DOI: 10.1201/9781003362258-6

(Mekonnen, 2021). Liposomes act as a carrier for radioisotopes and thus are used in diagnosis (Samia et al., 2003). Quantum dots, which are semiconductor nanocrystals possessing special properties, are highly sensitive to even low concentrations of analytes (Sobik et al., 2011).

Several diseases, both infectious and non-infectious, commonly affecting the majority of living beings, are now detected precisely by the use of nanotechnology. One example is cancer, which is highly prevalent among both the human and animal populations. Cancer is a huge burden to society and a challenge to public health worldwide, which is better identified using several nano-based detection methods, for example, nano-based molecular imaging and nano-based ultrasensitive biomarker detection, which act as promising tools to aid in the convenient and cost-effective diagnosis of cancer (Chen et al., 2018). Several biomarkers are used in the diagnosis of bladder cancer. These include gold nanoparticles, graphene, Galectin-1 protein, and many more (Barani et al., 2021). There have been positive results from the fusion of nanotechnology with artificial intelligence to supplement precision diagnostics (Adir et al., 2020). Fluorescence imaging, contrast-enhanced ultrasound (CEUS), positron emission tomography (PET), MRI, and CTA are nano-based diagnostic methods for atherosclerosis (Kratz et al., 2016). This further helps in timely diagnosing and treating coronary artery disease (Karimi et al., 2016). Owing to their unique properties, there has been an increase in attempts to evaluate nanotechnology-based systems in terms of their use in managing sepsis. Nanosensors can be used to detect sepsis biomarkers such as procalcitonin accurately. This can be attributed to their principles based on magnetism, immunology, and electrochemistry (Pant et al., 2021). Also, nanowires, cantilevers, and quantum dots have diagnostic applications in several gastrointestinal disorders (Laroui et al., 2013). Nanotechnology has also been a boon in the difficult times of the COVID-19 pandemic through its role in both diagnosis and treatment. SARS-COV-2 RNA targets and IgM are easily detected by AuNP-based calorimetric assay. AuNP-based LFAs can even quantify analytes. Optomagnetic biosensors, NP-based breathalyzer sensor assay, and fluorescence-linked immunoassay are some other diagnostic methods (Rashidzadeh et al., 2021). Nanosensors are miniature devices used to monitor important physiological parameters and diagnose tissue samples. To monitor animals, there are wearable technologies to detect body temperature, movement, stress, sweat composition, and antibiotics. Sensors monitor BHBA levels in animals to diagnose subclinical ketosis. Nanotubes can detect estrus, and nanosensors can detect genital infections in animals (Youssef et al., 2019).

6.2.2 NANOBIOTECHNOLOGY IN ANIMAL PRODUCTION

Nanobiotechnology is a rapidly expanding field and has potential applications in the veterinary field. Zinc, silver, copper, gold, selenium, and calcium nanoparticles are potential feed additives for animals (Mohd Yusof et al., 2021). Low doses of metal-containing nanoparticles will have the same effect on animal production as higher doses of mineral feed additives (Leng, 2018). Nanoparticle-based feed additives have been proven to improve hatchability and growth in the poultry industry (El-Rayes et al., 2019). They modulate intestinal flora and improve intestinal health (Feng et al.,

2017). They maintain blood homeostasis, prevent oxidative damage, and enhance the immune response thus enhancing productivity and growth, and improving carcass traits. They improve rumen fermentation and milk production in ruminants. Due to their nanoscale, they can increase the bioavailability of nutrients and hence are more beneficial and more eco-friendly than conventional methods (Mohd Yusof et al., 2021).

Zinc oxide nanoparticles are a promising alternative to traditional antibiotics due to their biocompatible nature and excellent antibacterial potential. They not only serve as an alternative antibacterial agent to control disease but also promote growth and performance (Mehdi et al., 2018). ZnO NPs exhibit unique properties such as extensive surface area, biocompatibility, biodegradability, semiconductor behavior, and UV light barrier that allows its use in a wide range of applications (Mohd Yusof et al., 2021).

There has been significant enhancement of cell-mediated immunity, in terms of mean skin thickness sensitive to phytohemagglutinin, in chicks given nano-silver treatment. Smart drug-delivery systems have the potential to manipulate structures or other particles at the nanoscale and control and catalyze chemical reactions and have self-regulating capabilities (Li et al., 2019). Nanotechnology is also capable of tracking estrus in animals because these tubes can detect estradiol antibodies by near-infrared fluorescence (Youssef et al., 2019). Nanotechnology has tremendous potential to revolutionize animal health and boost its production.

6.3   NANOTECHNOLOGY AND BIOSAFETY

6.3.1   ROLE OF NANOPARTICLES IN CREATING A STERILE ENVIRONMENT IN CONTAINMENT ZONES

Nanotechnology is an increasingly popular multi-facet technique that pertains to the engineering of practical networks at the molecular scale. Nanoparticles give new alternatives in integrating medicine and modern equipment, with the existing accomplishment simply commencing to scrape the surface of their numerous advantages and possible uses. The prevention of infectious agent transmission and epidemics is one of the main challenges to bioscience nowadays, requiring analysis undertakings to formulate nanoparticles for prophylactic antiviral applications. Nanoparticles give another significant application in the employment of inorganic nanoparticles in antiseptic sheets to inhibit infective agent transmission, which appears to be a promising development. The elevated surface area to volume ratio and distinctive chemical properties of metal-based nanoparticles enable their robust inactivation of viruses. Nanoparticles gain their virucidal action by mechanisms including inhibition of virus–cell receptor binding and reactive oxygen species chemical reaction (Lin et al., 2021).

There are a variety of possible nanoparticles that could impede infectious agent infections, notably in avoiding attachment, such as titanium oxide, zinc oxide, iron, silver, and gold, by manipulating the virus before infection and in the initial stages of its pathological process through self-sterilizing, which is an excellent strategy since essential infectious agent elements such as the spike proteins are accessible to degradation (Lara et al., 2010; Park et al., 2014; Meléndez-Villanueva et al., 2019).

Nanotechnology presents many opportunities for additional economical and promising disinfectant systems. It opens the way for a self-cleaning methodology. These systems will have antimicrobial activity or be able to release chemical disinfectants slowly, increasing their time of action (Hsu et al., 2011). Also, they will contribute extra properties, such as responsive systems, that deliver active substances in response to different stimuli, such as thermal and catalytic activity (Dalawai et al., 2020; Geyer et al., 2020). Some metallic nanoparticles also are better known for possessing a wide spectrum of activities against microorganisms acting as a broad-spectrum treatment (Dyshlyuk et al., 2020). Also, the engineered water nanostructures (EWNS) generated through an electrospray also possess antimicrobial activity. The results showed a large deduction in infective agent concentration (including H1N1 influenza). Additionally, the functional ingredient (hydrogen peroxide) doses required for inactivation were considerably lower (nanogram level), implying the viability of this platform (StatNano, 2020). Silver, copper, and zinc show intrinsic antipathogenic properties. It is advantageous to use NPs made of these metals instead of bulk equipment or the metal ions because of NPs releasing the cyanogenic metal ions slowly and increasingly wherever the antimicrobial action is required, and as NPs will deposit among cells while not being deported by specialized flow pumps (Sim et al., 2018). In addition to metal NPs, graphene derivatives have additionally shown promising infectious agent inactivation properties (Tu et al., 2018). For example, graphene oxide (GO) sheets and sulfated GO derivatives have been to be effective against herpes simplex virus type-1 (HSV-1) infections, with infectious agent binding and shielding because of the purported main repressive mechanisms (Sametband et al., 2014). Additionally, the antiviral activity of graphene derivatives has been associated with the negative surface charges and sharp edges of the individualized sheets, which can interfere with the binding of the virus to host cells (Song et al., 2015; Ye et al., 2015). Figure 6.2 shows antiviral effect of silver nanoparticles. The virucidal impact of an electrically charged disinfectant (CAC-717) that encounters mesostructure nanoparticles on engulfed viruses enabled the eradication of the infectious agent population (Nakashima et al., 2017). The paradigm transformations within the metallic crystals from large to minute-sized to nanoscale have resulted in excellent characteristics with outstanding properties for a broad variety of applications. Notably, AgNPs have gained a lot of interest as a result of their distinctive optical and chemical change properties which can be adapted by their surface nature and size ratio. The various nanomaterials are utilized as economical disinfectants by enhancing their physical properties (Deshmukh et al., 2019a).

6.4 NANOTECHNOLOGY AND VETERINARY CARE

6.4.1 ROLE OF NANOTECHNOLOGY IN MAINTAINING THE STERILITY OF VETERINARY IMPLANTS/CATHETERS

A catheter is an artificial tube-like structure that is inserted into the body cavity for the treatment of various diseases and performing surgical procedures. For proper drainage and to allow medicines to be administered via a catheter into the body, it is a prerequisite that it should maintain a sterile environment inside the body (Diggery and Grint, 2012). With recent advancements, various silver nanoparticles are coated over catheters to achieve antibacterial activity and thus prevent the formation of surface biofilm (Franci et al., 2015). Urinary catheters such as the Foley catheter are used in almost all operations related to urinary obstruction, urine diversion while performing surgery, and urolithiasis (Davis and Silverman, 2019). Nanomaterials such as carbon nanotubes are widely used in veterinary surgery because of their increased strength and low thrombogenic risk (Mali, 2013). Implants are devices used in medical and veterinary sciences so that any replacement or damage to a biological structure can be managed and its function can be enhanced (Yimin, 2016). The lifespan of implants in joint replacement surgery has been improved through the usage of nanoengineered implants. Osteoblastic cellular activity has been also improved with the use of nanotextured surfaces over surgical implants (Shi et al., 2010). Often, bacteria can adhere to surgical implants, resulting in periprosthetic joint infections. Nanophasic prosthetic joints are among the best ways to solve such major problems (Gavaskar et al., 2018). In addition, porous scaffolds and hydrogels are used along with the incorporation of growth factors and polymeric nanoparticles. Cell sheet engineering strategies for the regeneration of novel tissues are made possible by the use of nano-technology (Shi et al., 2010). Zinc oxide, gold, copper, silver, and other nanoparticles can be used for wound healing and the coating of medical implants and devices. Nanofibers used in scaffolds help in adhesion, cell proliferation, and differentiation, and subsequent tissue formation can be judged by various nanosensors. Various graft implants for cardiovascular and respiratory systems are made using polymer polyhedral oligomeric silsesquioxane poly (carbonate urea) urethane (POSS-PCU). Nano-antimicrobials are added to hydrogels to enable advancements in ocular sur-gery, thus obtaining tough and flexible properties in corneal implants (Agnihotri and Dhiman, 2017).

6.4.2   NANOPARTICLES AS DISINFECTANTS FOR LABORATORY EQUIPMENT

Disinfectants are chemicals that can remove almost all pathogenic microorganisms from inanimate/non-living objects (Mohapatra, 2017). Commonly used disinfectants include phenol, hydrogen peroxide, etc., but, due to their corrosive nature, there are limitations to using them as disinfectants for lab equipment. Nanotechnology has entered this field also due to the antimicrobial action of silver and its compounds, and it has emerged as a potent disinfectant for cleansing laboratory equipment. Silver polyamide has an antibacterial effect through the sustainable release of silver ions over the area in which it is used (Deshmukh et al., 2019a). The antibacterial and anti-viral effects of silver nanoparticles are shown in Figures 6.1 and 6.2, respectively.

Various disinfectants used for laboratory equipment include the following:

- The formulation of titanium dioxide and silver nanoparticles provided self-sterilization to the surface and was used widely in the COVID-19 pandemic in many countries. After 2 minutes of treatment, the mesostructure nanoparticles

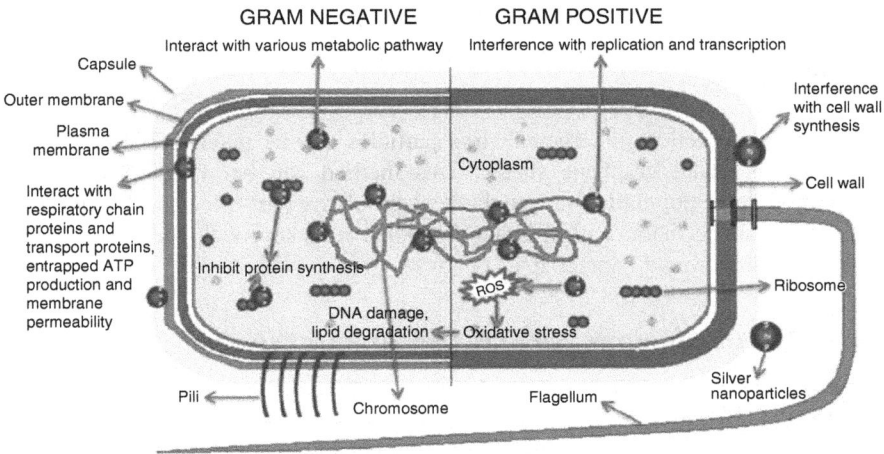

FIGURE 6.1 Antibacterial mechanism of silver nanoparticles.

Source: Figure adapted from Deshmukh et al., 2019 after permission.

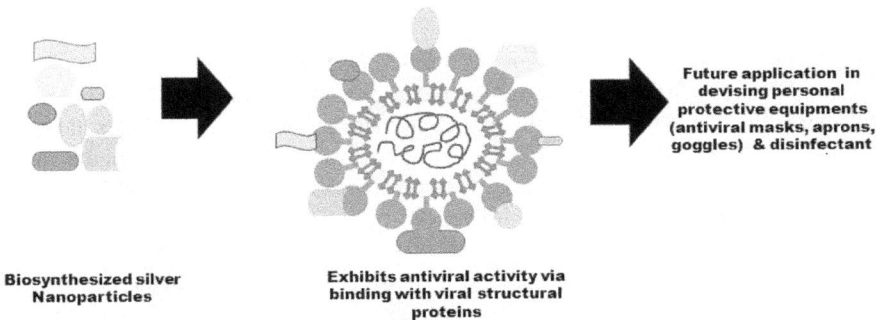

FIGURE 6.2 Future prospects of silver nanoparticles as antivirals.

having the matrix of electrically charged disinfectant (CAC-717) help in reducing the viral load to below the detection limit.

- The inhibitory effect shown by Ag_3O-SiO_2 particles is used for the prevention of influenza A virus.
- The photocatalytic nanostructured film having a matrix of titanium dioxide and silicon has a high disinfectant power. When the UV-A activated system is exposed to it for 20 minutes, it causes great damage to microorganisms by interacting with the membrane.
- Quaternary ammonium salt has a long-lasting durable action and possesses high sterilization capacity.
- A nanofilm of sodium chlorite ($NaClO_2$) is a widely used disinfectant due to the liberation of disinfectant gas (ClO_2) after it is activated by UV rays and exposed to moisture.

- Cellulose nanostructures, silica nanocomposites, iron and silver biogenic nanoparticles, etc., can be used as a disinfectant for laboratory equipment (Campos et al., 2020).

The rapid detection of targeted virus particles can be made possible by using antibody-conjugated graphene sheets. This method can be widely used for the screening of large populations as graphene materials are cost-effective. Many sensors and filters can be made with graphene nanoparticles to check for environmental pollution (Palmieri and Papi, 2020).

6.4.3 NANOTECHNOLOGY AND ANIMAL HEALTHCARE WORKERS

Research statistics from the Food and Agriculture Organization of the United Nations estimate that livestock resources contribute up to 40% of total agricultural output in developed countries and 20% in developing countries (Food and Agriculture Organization of the United Nations, 2020). In such a scenario, animal health workers have always stepped up to provide preventive health care and first aid care to animals and poultry, and also contribute to the prevention and control of various contagious diseases. Thus, it becomes crucial that the health of animal care workers is given extreme importance too.

Over the last two decades, the world has witnessed the emergence of nanotechnology and its myriad applications in numerous fields. One of the most widely researched areas for the implementation of nanotechnology is the field of healthcare and medicine. Nanomedicine is the branch of nanotechnology that utilizes nanotechnology and its allied nanosystems in medicine to achieve various benefits in disease prevention, diagnosis, and treatment (Anjum et al., 2021). Apart from these, the health of an individual is also influenced by their environment. In this context, nanotechnology has presented numerous promising avenues for reducing environmental pollution and ensuring clean water, fresh air, and good health for humankind. The high surface area-to-volume ratio results in higher reactivity and enhanced properties combined with reduced costs of nanomaterials, making them considerably appropriate for ameliorating both the environment and health (Guerra et al., 2018). Additionally, nanotechnology has also proved its mettle in the field of food safety, from production to packaging, thus scaling down the incidences of various food borne illnesses and poisonings (Sekhon, 2010). The application of bionanotechnology in the food industry is shown in Figure 6.3.

The following section explores the various aspects where nanotechnology has proved to be very effective in creating a clean and green environment for better health:

- Nanotechnology for purification of water and air (Mandeep and Shukla, 2020; Gehrke et al., 2015).
- Nanotechnology and degradation of hazardous chemicals for their safe disposal (Kumar Singh et al., 2022).
- Nanotechnology and food safety (Singh et al., 2017).

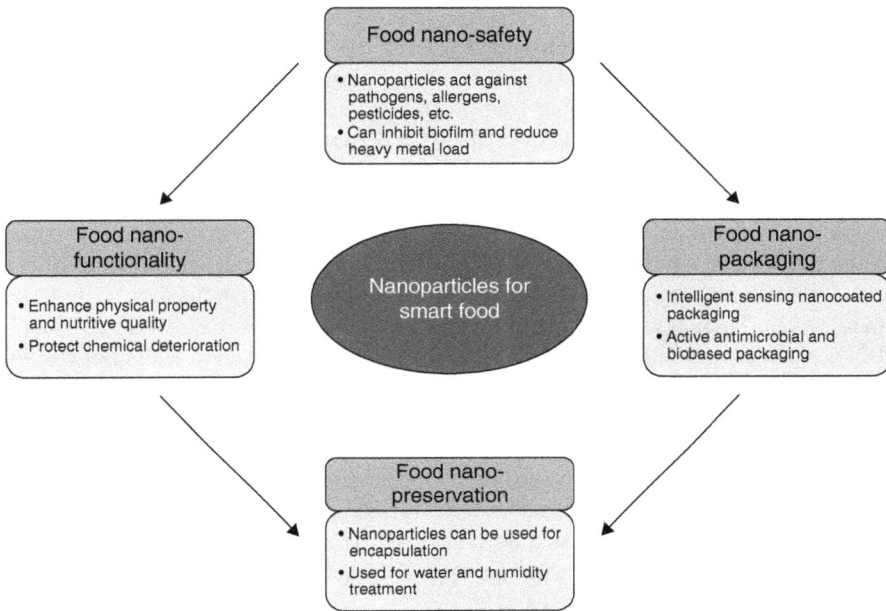

FIGURE 6.3 Application of bionanotechnology in the food industry.

6.4.4 NANOTECHNOLOGY IN MAINTAINING A STERILE ENVIRONMENT FOR LIVESTOCK CARE

Infectious diseases related to livestock are a major concern as they are directly linked with human food production. In addition to conventional methods, newer strategies such as the use of nanobiotechnology are being encouraged in maintaining a sterile environment for livestock care. Disinfectants are widely used as efficient, cheaper antimicrobial agents but they can have drawbacks such as being corrosive, harmful, and resistant to development (Deshmukh et al., 2019b). Silver nanoparticles are used as a surface disinfectant, water disinfectant, and therapeutic material as they are more potent and show a broader spectrum of action due to their physiochemical properties such as larger surface-to-volume ratio, shape, size, and phase (Pandey et al., 2014). Their bactericidal properties have been proved to be effective against *Staphylococcus aureus*, *Pseudomonas aeruginosa*, *Escherichia coli*, *Bacillus cereus*, *Listeria innocua*, and *Salmonella cholera suis* (Pantic, 2014). They do not only inhibit the reproduction of bacteria but also of viruses and fungi. They are effective against *Aspergillum*, *Candida*, and *Saccharomyces*. Replication of the H1V1 virus is inhibited by Ag nanoparticles without showing harmful effects on humans (Sun et al., 2005). Silver nanoparticles play an important role in the prevention of contamination in animal breeding facilities and transportation chambers (Chmielowiec-Korzeniowska et al., 2015).

Various prevalent diseases including foot-and-mouth disease (FMD), brucellosis, Johne's disease, anaplasmosis, and bovine respiratory disease complex (BRDC) cause huge economic and production losses. Zoonotic diseases are transmitted via milk, meat, and various other animal products, therefore, it is essential to stop the spread of disease and prevent contamination of the environment (Maina et al., 2020). Nano-vaccines have an advantage over traditional vaccines as they do not require strict storage conditions (Wagner-Muñiz et al., 2018). They induce rapid and long-lasting immunity and do not require booster doses (Ulery et al., 2011; Wafa et al., 2017).

The toxicity shown by silver nanoparticles is a major risk factor as it can cause bio-accumulation and can lead to organ failure in animals and also in humans consuming them over the long term(Ahamed et al., 2010). They cannot be removed easily by common methods such as rinsing and continue to cause unpredictable toxicity in the environment. There is also the risk of the development of microbial resistance (Saadh, 2021). A major obstacle is minimal research on the use of nanobiotechnology in the livestock sector.

Nanoparticles have significant potential and have many advantages over trad-itional practices, and the feasibility of industrial scale-up is of utmost importance (Wagner-Muñiz et al., 2018). Spray-drying approaches in nano-vaccines are being considered to minimize the prerequisite of further product purification and ensure more uniform batches (Sosnik and Seremeta, 2015). In the future, there will be pro-gressive advances in nonbiomedical science and its use in improving animal health, hence much research is urgently needed.

6.5 NANOTECHNOLOGY AND THE DEGRADATION OF HAZARDOUS CHEMICALS FOR THEIR SAFE DISPOSAL

The current age of technology and industrialization has paved the way for enor-mous quantities of hazardous waste, whose safe disposal poses a major obstacle. Hazardous and toxic materials are the byproducts of many industries such as manu-facturing, agriculture, hospitals, construction, etc. They may include a wide range of chemicals, radiation, heavy metals, and even pathogens. Some common examples include paints, pesticides, clinical waste, persistent organic pollutants (POPs), strong acids and alkalis, arsenic, lead, cadmium, etc. The progress and research in the field of nanotechnology have put forward a few promising methods by which the safe dis-posal of some of these hazardous chemicals is possible. Nanomaterials, due to their high specific surface areas, could be very effectively used as catalysts, adsorbents, membranes, or additives to reduce or transform waste materials (Khan and Lee, 2021).

- Photocatalysis using metal oxide semiconductor nanostructures has been shown to be effective in degrading organic contaminants in water. They accel-erate photoreactions without undergoing any physicochemical alterations them-selves (Baruah et al., 2008; Baruah and Dutta, 2009). Photo-electrocatalytic dye degradation is one of the techniques inviting substantial interest in the deg-radation and safe disposal of chemical dye-containing wastewater (Khan and Lee, 2021).
- Eco-friendly remediation of pollutants is possible through the fabrication of "green nanomaterials" from microorganisms and extracts of other organisms

(Mandeep and Shukla, 2020). Iron nanoparticles are among the most effective green nanoparticles for remediation, owing to their redox potential on reaction with water, magnetic susceptibility, and non-toxic nature (Bolade et al., 2020).

- Nano-adsorbents are adept at removing both organic and inorganic pollutants. They are classified chiefly into metal-based, carbon-based, and metal oxide-based nanoparticles. Carbon nanotubes (CNTs), graphene, activated carbon, and fullerene are some of the carbon-based nanoparticles in use. Their role as adsorbents for toxic chemicals in manufacturing and pharmaceutical industries is very efficacious (Kumari et al., 2019).
- Activated carbon-modified nano-magnets have proved to be very efficacious in eliminating fluoride ions from wastewater. With an uptake of 454.54 mg/g, the nanocomposite was described as removing 97.4% of fluoride ions from synthetic wastewater by the process of sorption (Takmil et al., 2020).
- Researchers all over the world are focused on altering waste materials in such a manner as to change their properties. This approach will help to reduce waste and generate useful products concurrently. This perspective can be effectively used in the production of biogas, biohydrogen, adsorbents, clinker, biomolecules, and many other products in the industrial sector (Mandeep et al., 2020). Nanotechnology has proved to be very promising in the enhancing the rate of production, and ensuring the efficient transformation of waste materials into useful resources. For instance, the use of nanoparticles to enhance dark fermentation reactions, resulting in increased biohydrogen production has been successfully described (Kumar et al., 2019). Additionally, supplementation of fermentative bacteria with nanoparticles has created new routes for biohydrogen generation from wastewater (Elreedy et al., 2019).

6.6   CONCLUSION

Over the past few years, the application of nanotechnology in human and veterinary medicine has shown significant progress. It has created high expectations in animal health, veterinary medicine, and other areas of animal production. Looking into the vast areas of application of nanotechnology in diagnostics, disease treatment, vaccinology, creating sterile environments for health care workers, biosafety, and biowaste disposal, nanotechnology holds high promise for the next generation. In veterinary sciences, nanotechnology has already spread to the domains of diagnostics and vaccinology. More research is on-going around the world on nanobiotechnology to improve the quality of life of animals and humans. In the forthcoming years, nanotechnology research will reform the science and technology of animal health and will help to boost livestock production. Nanotechnology in veterinary sciences has the potential to improve diagnosis and treatment, and also to provide a better environment for healthcare workers.

REFERENCES

Adir, O., Poley, M., Chen, G., Froim, S., Krinsky, N., Shklover, J., Shainsky Roitman, J., Lammers, T., Schroeder, A. (2020) Integrating artificial intelligence and nanotechnology for precision cancer medicine. *Advanced Materials* 32: 1901989.

Agnihotri, S., Dhiman, N.K. (2017) Development of nano-antimicrobial biomaterials for bio-medical applications. In: A. Tripathi, J.S. Melo (Eds.) *Advances in Biomaterials for Biomedical Applications*, pp. 479–545. Springer Singapore: Singapore.

Ahamed, M., AlSalhi, M.S., Siddiqui, M.K.J. (2010) Silver nanoparticle applications and human health. *Clinica Chimica Acta* 411: 1841–1848.

Anjum, S., Ishaque, S., Fatima, H., Farooq, W., Hano, C., Abbasi, B.H., Anjum, I. (2021) Emerging applications of nanotechnology in healthcare systems: Grand challenges and perspectives. *Pharmaceuticals* 14: 707.

Barani, M., Hosseinikhah, S.M., Rahdar, A., Farhoudi, L., Arshad, R., Cucchiarini, M., Pandey, S. (2021) Nanotechnology in bladder cancer: Diagnosis and treatment. *Cancers* 13: 2214.

Baruah, S., Dutta, J. (2009) Nanotechnology applications in pollution sensing and degradation in agriculture: a review. *Environmental Chemistry Letters* 7: 191–204.

Baruah, S., Rafique, R.F., Dutta, J. (2008) Visible light photocatalysis by tailoring crystal defects in zinc oxide nanostructures. *Nano* 3: 399–407.

Bolade, O.P., Williams, A.B., Benson, N.U. (2020) Green synthesis of iron-based nanomaterials for environmental remediation: A review. *Environmental Nanotechnology, Monitoring & Management* 13: 100279.

Campos, E.V.R., Pereira, A.E.S., de Oliveira, J.L., Carvalho, L.B., Guilger-Casagrande, M., de Lima, R., Fraceto, L.F. (2020) How can nanotechnology help to combat COVID-19? Opportunities and urgent need. *Journal of Nanobiotechnology* 18: 125.

Chen, X.-J., Zhang, X.-Q., Liu, Q., Zhang, J., Zhou, G. (2018) Nanotechnology: a promising method for oral cancer detection and diagnosis. *Journal of Nanobiotechnology* 16: 52.

Chmielowiec-Korzeniowska, A., Tymczyna, L., Dobrowolska, M., Banach, M., Nowakowicz-Dębek, B., Bryl, M., Drabik, A., Tymczyna-Sobotka, M., Kolejko, M. (2015) Silver (Ag) in tissues and eggshells, biochemical parameters and oxidative stress in chickens. *Open Chemistry* 13: 000010151520150140.

Dalawai, S.P., Aly, M.A.S., Latthe, S.S., Xing, R., Sutar, R.S., Nagappan, S., Ha, C.-S., Sadasivuni, K., kumar, Liu, S. (2020) Recent advances in durability of superhydrophobic self-cleaning technology: A critical review. *Progress in Organic Coatings* 138: 105381.

Davis, J.E., Silverman, M.A. (2019) Urologic procedures. In: T.W. Thomsen (Ed.) *Roberts and Hedges' Clinical Procedures in Emergency Medicine and Acute Care.* Elsevier: Philadelphia, PA.

Deshmukh, S.P., Patil, S.M., Mullani, S.B., Delekar, S.D. (2019a) Silver nanoparticles as an effective disinfectant: A review. *Materials Science and Engineering: C* 97: 954–965.

Diggery, R.C., Grint, D.T. (Eds.) (2012) *Catheters: Types, Applications, and Potential Complications.* Nova Science: Hauppauge, NY.

Dyshlyuk, L.S., Babich, O., Ivanova, S., Vasilchenco, N., Prosekov, A.Y., Sukhikh, S. (2020) Suspensions of metal nanoparticles as a basis for protection of internal surfaces of building structures from biodegradation. *Case Studies in Construction Materials* 12: e00319.

El-Rayes, T., El-Damrawy, S., El-Deeb, M., Adel Abdelghany, I. (2019) Re/post-hatch nano-zinc supplementations effects on hatchability, growth performance, carcass traits, bone characteristics and physiological status of inshas chicks. *Egyptian Poultry Science Journal* 39: 771–789.

Elreedy, A., Fujii, M., Koyama, M., Nakasaki, K., Tawfik, A. (2019) Enhanced fermenta-tive hydrogen production from industrial wastewater using mixed culture bacteria incorporated with iron, nickel, and zinc-based nanoparticles. *Water Research* 151: 349–361.

Feng, Y., Min, L., Zhang, W., Liu, J., Hou, Z., Chu, M., Li, L., Shen, W., Zhao, Y., Zhang, H. (2017) Zinc oxide nanoparticles influence microflora in ileal digesta and correlate well with blood metabolites. *Frontiers in Microbiology* 8.

Food and Agriculture Organization of the United Nations (2020) *World Food and Agriculture–Statistical Yearbook 2020*. Food and Agriculture Organization of the United Nations: Rome, Italy.

Franci, G., Falanga, A., Galdiero, S., Palomba, L., Rai, M., Morelli, G., Galdiero, M. (2015) Silver nanoparticles as potential antibacterial agents. *Molecules* 20: 8856–8874.

Gavaskar, A., Rojas, D., Videla, F. (2018) Nanotechnology: the scope and potential applications in orthopedic surgery. *European Journal of Orthopaedic Surgery & Traumatology* 28: 1257–1260.

Gehrke, I., Geiser, A., Somborn-Schulz, A. (2015) Innovations in nanotechnology for water treatment. *Nanotechnology, Science and Applications* 1.

Geyer, F., D'Acunzi, M., Sharifi-Aghili, A., Saal, A., Gao, N., Kaltbeitzel, A., Sloot, T.-F., Berger, R., Butt, H.-J., Vollmer, D. (2020) When and how self-cleaning of superhydrophobic surfaces works. *Science Advances* 6: eaaw9727.

Guerra, F., Attia, M., Whitehead, D., Alexis, F. (2018) Nanotechnology for environmental remediation: Materials and applications. *Molecules* 23: 1760.

Hsu, B.B., Wong, S.Y., Hammond, P.T., Chen, J., Klibanov, A.M. (2011) Mechanism of inactivation of influenza viruses by immobilized hydrophobic polycations. *Proceedings of the National Academy of Sciences* 108: 61–66.

Jo, S.H., Chang, T., Ebong, I., Bhadviya, B.B., Mazumder, P., Lu, W. (2010) Nanoscale memristor device as synapse in neuromorphic systems. *Nano Letters* 10: 1297–1301.

Karimi, M., Zare, H., Bakhshian Nik, A., Yazdani, N., Hamrang, M., Mohamed, E., SahandiZangabad, P., Moosavi Basri, S.M., Bakhtiari, L., Hamblin, M.R. (2016) Nanotechnology in diagnosis and treatment of coronary artery disease. *Nanomedicine* 11: 513–530.

Khan, S.B., Lee, S.L. (2021) Nanomaterials significance; contaminants degradation for environmental applications. *Nano Express* 2: 022002.

Kratz, J.D., Chaddha, A., Bhattacharjee, S., Goonewardena, S.N. (2016) Atherosclerosis and nanotechnology: Diagnostic and therapeutic applications. *Cardiovascular Drugs and Therapy* 30: 33–39.

Kumar, M., Gupta, G., Shukla, P. (2020) Insights into the resources generation from pulp and paper industry wastes: Challenges, perspectives and innovations. *Bioresource Technology* 297: 122496.

Kumar, G., Mathimani, T., Rene, E.R., Pugazhendhi, A. (2019) Application of nanotechnology in dark fermentation for enhanced biohydrogen production using inorganic nanoparticles. *International Journal of Hydrogen Energy* 44: 13106–13113.

Kumar Singh, S., Singh, S., Singh Gautam, A., Kumar, V., Singh Rajput, R., Singh Rajput, M. (2022) Nanoparticles: Novel approach to mitigate environmental pollutants. In: K. Ferreira Mendes, R. Nogueira de Sousa, K. Cabral Mielke (Eds.) *Biodegradation Technology of Organic and Inorganic Pollutants*. Intech Open.

Kumari, P., Alam, M., Siddiqi, W.A. (2019) Usage of nanoparticles as adsorbents for waste water treatment: An emerging trend. *Sustainable Materials and Technologies* 22: e00128.

Lara, H.H., Ayala-Núñez, N.V., IxtepanTurrent, L. del C., Rodríguez Padilla, C. (2010) Bactericidal effect of silver nanoparticles against multidrug-resistant bacteria. *World Journal of Microbiology and Biotechnology* 26: 615–621.

Laroui, H., Rakhya, P., Xiao, B., Viennois, E., Merlin, D. (2013) Nanotechnology in diagnostics and therapeutics for gastrointestinal disorders. *Digestive and Liver Disease* 45: 995–1002.

Leng, X.P.D. (2018) Effects of dietary zinc oxide nanoparticles supplementation on growth performance, zinc status, intestinal morphology, microflora population, and immune response in weaned pigs. *Journal of the Science of Food and Agriculture* 99: 1366–1374.

Li, C., Wang, J., Wang, Y., Gao, H., Wei, G., Huang, Y., Yu, H., Gan, Y., Wang, Y., Mei, L., Chen, H., Hu, H., Zhang, Z., Jin, Y. (2019) Recent progress in drug delivery. *Acta Pharmaceutica Sinica B* 9: 1145–1162.

Lin, N., Verma, D., Saini, N., Arbi, R., Munir, M., Jovic, M., Turak, A. (2021) Antiviral nanoparticles for sanitizing surfaces: A roadmap to self-sterilizing against COVID-19. *Nano Today* 40: 101267.

Maina, T.W., Grego, E.A., Boggiatto, P.M., Sacco, R.E., Narasimhan, B., McGill, J.L. (2020) Applications of nanovaccines for disease prevention in cattle. *Frontiers in Bioengineering and Biotechnology* 8: 608050.

Mali, S. (2013) Nanotechnology for surgeons. *Indian Journal of Surgery* 75: 485–492.

Mandeep, Shukla, P. (2020) Microbial nanotechnology for bioremediation of industrial wastewater. *Frontiers in Microbiology* 11: 590631.

Mehdi, Y., Létourneau-Montminy, M.-P., Gaucher, M.-L., Chorfi, Y., Suresh, G., Rouissi, T., Brar, S.K., Côté, C., Ramirez, A.A., Godbout, S. (2018) Use of antibiotics in broiler production: Global impacts and alternatives. *Animal Nutrition (Zhongguo Xu Mu Shou Yi Xue Hui)* 4: 170–178.

Mekonnen, G. (2021) Review on application of nanotechnology in animal health and production. *Journal of Nanomedicine and Nanotechnology* 12: 559.

Meléndez-Villanueva, M.A., Morán-Santibañez, K., Martínez-Sanmiguel, J.J., Rangel-López, R., Garza-Navarro, M.A., Rodríguez-Padilla, C., Zarate-Triviño, D.G., Trejo-Ávila, L.M. (2019) Virucidal activity of gold nanoparticles synthesized by green chemistry using garlic extract. *Viruses* 11: 1111.

Mohapatra, S. (2017) Sterilization and disinfection. In: *Essentials of Neuroanesthesia*, pp. 929–944. Elsevier.

Mohd Yusof, H., Abdul Rahman, N., Mohamad, R., Hasanah Zaidan, U., Samsudin, A.A. (2021) Antibacterial potential of biosynthesized zinc oxide nanoparticles against poultry-associated foodborne pathogens: An in vitro study. *Animals* 11: 2093.

Nakashima, R., Kawamoto, M., Miyazaki, S., Onishi, R., Furusaki, K., Osaki, M., Kirisawa, R., Sakudo, A., Onodera, T. (2017) Evaluation of calcium hydrogen carbonate mesoscopic crystals as a disinfectant for influenza A viruses. *Journal of Veterinary Medical Science* 79: 939–942.

Palmieri, V., Papi, M. (2020) Can graphene take part in the fight against COVID-19? *Nano Today* 33: 100883.

Pandey, J., Swarnkar, R., KK, S., Dwivedi, P., Singh, M., Sundaram, S., Gopal, R. (2014) Silver nanoparticles synthesized by pulsed laser ablation: as a potent antibacterial agent for human enteropathogenic Gram-positive and Gram-negative bacterial strains. *Applied Biochemistry and Biotechnology* 174(3): 1021–1031.

Pant, A., Mackraj, I., Govender, T. (2021) Advances in sepsis diagnosis and management: a paradigm shift towards nanotechnology. *Journal of Biomedical Science* 28: 6.

Pantic, I. (2014) Application of silver nanoparticles in experimental physiology and clinical medicine: current status and future prospects. *Reviews on Advanced Materials Science* 37.

Park, G.W., Cho, M., Cates, E.L., Lee, D., Oh, B.-T., Vinjé, J., Kim, J.-H. (2014) Fluorinated TiO_2 as an ambient light-activated virucidal surface coating material for the control of human norovirus. *Journal of Photochemistry and Photobiology B: Biology* 140: 315–320.

Rashidzadeh, H., Danafar, H., Rahimi, H., Mozafari, F., Salehiabar, M., Rahmati, M.A., Rahamooz-Haghighi, S., Mousazadeh, N., Mohammadi, A., Ertas, Y.N., Ramazani, A., Huseynova, I., Khalilov, R., Davaran, S., Webster, T.J., Kavetskyy, T., Eftekhari, A., Nosrati, H., Mirsaeidi, M. (2021) Nanotechnology against the novel coronavirus (severe acute respiratory syndrome coronavirus 2): diagnosis, treatment, therapy and future perspectives. *Nanomedicine* 16: 497–516.

Saadh, M.J. (2021) Synthesis, role in antibacterial, antiviral, and hazardous effects of silver nanoparticles. *Pharmacology Online* 2: 1331–1336.

Sametband, M., Kalt, I., Gedanken, A., Sarid, R. (2014) Herpes simplex virus type-1 attachment inhibition by functionalized graphene oxide. *ACS Applied Materials & Interfaces* 6: 1228–1235.

Samia, A.C.S., Chen, X., Burda, C. (2003) Semiconductor quantum dots for photodynamic therapy. *Journal of the American Chemical Society* 125: 15736–15737.

Savaliya, R., Shah, D., Singh, R., Kumar, A., Shanker, R., Dhawan, A., Singh, S. (2015) Nanotechnology in disease diagnostic techniques. *Current Drug Metabolism* 16: 645–661.

Sekhon, B.S. (2010) Food nanotechnology–an overview. *Nanotechnology, Science and Applications* 3: 1–15.

Shi, J., Votruba, A.R., Farokhzad, O.C., Langer, R. (2010) Nanotechnology in drug delivery and tissue engineering: From discovery to applications. *Nano Letters* 10: 3223–3230.

Sim, W., Barnard, R.T., Blaskovich, M.a.T., Ziora, Z.M. (2018) Antimicrobial silver in medicinal and consumer applications: A patent review of the past decade (2007–2017). *Antibiotics (Basel, Switzerland)* 7: E93.

Singh, T., Shukla, S., Kumar, P., Wahla, V., Bajpai, V.K., Rather, I.A. (2017) Application of nanotechnology in food science: Perception and overview. *Frontiers in Microbiology* 8: 1501.

Sobik, M., Pondman, K.M., Erné, B., Kuipers, B., Haken, B. ten, Rogalla, H. (2011) Magnetic nanoparticles for diagnosis and medical therapy. In: R. Klingeler, R.B. Sim (Eds.) *Carbon Nanotubes for Biomedical Applications*, pp. 85–95. Springer: Berlin, Heidelberg.

Song, Z., Wang, X., Zhu, G., Nian, Q., Zhou, H., Yang, D., Qin, C., Tang, R. (2015) Virus capture and destruction by label-free graphene oxide for detection and disinfection applications. *Small* 11: 1171–1176.

Sosnik, A., Seremeta, K.P. (2015) Advantages and challenges of the spray-drying technology for the production of pure drug particles and drug-loaded polymeric carriers. *Advances in Colloid and Interface Science* 223: 40–54.

StatNano (2020) Coronavirus: nanotech surface sanitizes Milan with nanomaterials remaining self-sterilized for years | STATNANO. 2020. https://statnano.com//news/67531/Coro navirus-Nanotech-Surface-Sanitizes-Milan-with-Nanomaterials-Remaining-Self-sterili zed-for-Years. Accessed 22 Oct 2020.

Sun, R.W.-Y., Chen, R., Chung, N.P.-Y., Ho, C.-M., Lin, C.-L.S., Che, C.-M. (2005) Silver nanoparticles fabricated in HEPES buffer exhibit cytoprotective activities toward HIV-1 infected cells. *Chemical Communications* 5059–5061.

Takmil, F., Esmaeili, H., Mousavi, S.M., Hashemi, S.A. (2020) Nano-magnetically modified activated carbon prepared by oak shell for treatment of wastewater containing fluoride ion. *Advanced Powder Technology* 31: 3236–3245.

Tu, Z., Guday, G., Adeli, M., Haag, R. (2018) Multivalent interactions between 2D nanomaterials and biointerfaces. *Advanced Materials* 30: 1706709.

Ulery, B.D., Nair, L.S., Laurencin, C.T. (2011) Biomedical applications of biodegradable polymers. *Journal of Polymer Science Part B: Polymer Physics* 49: 832–864.

Wafa, E.I., Geary, S.M., Goodman, J.T., Narasimhan, B., Salem, A.K. (2017) The effect of polyanhydride chemistry in particle-based cancer vaccines on the magnitude of the anti-tumor immune response. *Acta Biomaterialia* 50: 417–427.

Wagner-Muñiz, D.A., Haughney, S.L., Kelly, S.M., Wannemuehler, M.J., Narasimhan, B. (2018)Room temperature stable PspA-based nanovaccine induces protective immunity. *Frontiers in Immunology* 9: 325

Ye, S., Shao, K., Li, Z., Guo, N., Zuo, Y., Li, Q., Lu, Z., Chen, L., He, Q., Han, H. (2015) Antiviral activity of graphene oxide: How sharp edged structure and charge matter. *ACS Applied Materials & Interfaces* 7: 21571–21579.

Yimin, Q. (2016) Textiles for implants and regenerative medicine. In: *Medical Textile Materials*, pp. 133–143. Elsevier.

Youssef, F.S., El-Banna, H.A., Elzorba, H.Y., Galal, A.M. (2019) Application of some nanoparticles in the field of veterinary medicine. *International Journal of Veterinary Science and Medicine* 7: 78–93.

7 Bionanotechnology in Food and Cosmetics

Brij Mohan, Jude Juventus Aweya,
Quansheng Chen, and Ekta Poonia

7.1 INTRODUCTION

Biotechnology is the branch of the science dealing with nanoscale materials and engineering functional molecular systems. Nanotechnology provides an active platform with broad applications in various sectors, including textiles, paper, food, concrete, agriculture, transport, energy, and medicine. Nanomaterials can be composited with biomolecules to enhance their activity and usefulness in various fields [1]. With the increasing global population, various public areas are growing very fast. This has caused different types of social, medical, and environmental safety issues. Primarily, everyone needs healthy and safe food. Another commonly used product in the home is cosmetic products that should be safe [2–4].

Human daily activities and needs impact health and safety. Therefore, safe food and safe products in everyday life are most important. Any toxic compounds used and/or consumed can create severe issues for health and vitality. Food toxicity can be due to bacteria, viruses, pesticides, heavy metal ions, etc. It can cause serious effects such as illness, vomiting, abdominal pain, kidney failure, heart diseases, etc. In addition, the toxicity of cosmetic compounds can be responsible for skin and lung cancers, and heart disease. Food toxicity can be due to mycotoxins, pesticide residues, process-induced toxins and drugs, bacteria and viruses, and allergens. Also, cosmetic products contaminated with lead, cadmium, mercury, chromium, nickel, arsenic, hydroquinone, steroids, nitrosamine, etc., are harmful to humans (Figure 7.1) [5,6].

Various materials and molecules have recently been developed to monitor the toxicity of food and cosmetic products. Materials such as metal-organic frameworks, nanomaterials, aptasensors, biomolecule-functionalized nanomaterials, etc. [7–10] have shown excellent applications in the real-time monitoring of toxic compounds in food and cosmetic products. Bionanomaterials have gained considerable attention due to their easy synthesis and high sensitivity. Moreover, spectroscopic and electrochemical nondestructive methods make toxicity monitoring easy. Therefore, it provides a cost-effective and easy-handling platform for researchers [11,12].

DOI: 10.1201/9781003362258-7

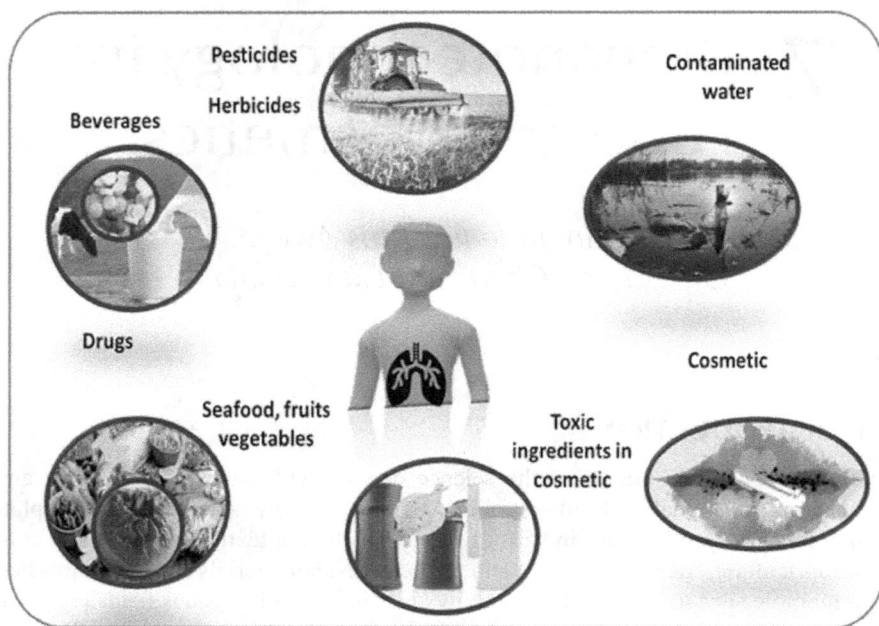

FIGURE 7.1   Food and cosmetic contaminants.

7.2   SCOPE AND APPROACH

This chapter discusses the analyte-sensing properties of bionanomaterials from food and cosmetic products. First, the development of materials, properties, and advantages of their application in biosensing are discussed in depth. Then, the current progress in bionanomaterials for food and cosmetic monitoring is discussed. Finally, the challenges and future directions are discussed for the design of materials with enhanced properties such as excellent bioactivities, rapid responsiveness, low cost, biodegradability, and non-toxicity.

7.3   BIONANOTECHNOLOGY-BASED MATERIALS

The incorporation of biomolecules into nano-artifacts is termed bionanotechnology [13]. The molecular species with multifunctional sites interacting with nanomaterials enable the assembly of unique structures. These unique structures and active sites exhibit the capability to capture toxic chemicals from products. In addition, the choice of precursors with charging site can enhance signal transductions [14,15].

Nanobiotechnology is an emerging area in the food and cosmetic sectors. Various nanostructures have been designed for the sensitive detection of analytes from target samples [16]. Moreover, different NPs, including gold, silver, platinum, zinc, cerium, iron, titanium dioxide, and quantum dots, have provided an active platform for capturing analytes through electrostatic interactions. Moreover, the biological properties such as antibacterial, antifungal, and antipathogenic of these NPs are appreciable [17,18].

The conventional methods suffer from sensitivity, selectivity, and capacity limitations. Bionanotechnology has provided a new and advanced platform with high sensitivity that can be used for multiple purposes. Moreover, the qualitative and quantitative determination of analytes is most important. Therefore, bionanotechnology-based biosensors are promising candidates for ultra-low-level detection and clarifying the analytical parameters [19].

7.4   DESIGN OF BIONANOTECHNOLOGY-BASED BIOSENSORS

Nanomaterials can be designed by carbon materials and biomolecules using layering methods, solvothermal, hydrothermal, microwave-assisted, mechanochemical, electrochemical, sonochemical synthesis, etc. The designed nanomaterials with biomolecules exhibit an active platform enriched with multifunctional sites and increased surface area. Moreover, immobilizing biomolecules leads to advanced materials having active sites and surfaces. Metals (Ag, Au, Pt, Ru, Pd, etc.) are commonly used for NP synthesis. In addition, quantum dots (QDs) and carbon nanotubes (CNTs) have been used as promising materials for designing nanomaterials. Moreover, deoxyribonucleic acid (DNA), amino acids, nucleic acids, and other small biomolecules are used to immobilize and create bionanomaterials (Figure 7.2) [20,21]. Therefore, bionanomaterial applications are explored by maximizing signal-generating sites and have been widely used in catalytic, sensing, drug delivery, adsorption, transport, etc. [22]

The synergetic effect of bionanomaterials with activated surfaces is appreciable for advanced sensing applications [23]. The increasing need for food and cosmetics has created a significant safety concern. Bionanomaterials for detecting analytes from food and cosmetics lead to a safety monitoring platform. Moreover, the fabricated biomolecules with metal- and carbon-based nanomaterials prevent leaching and

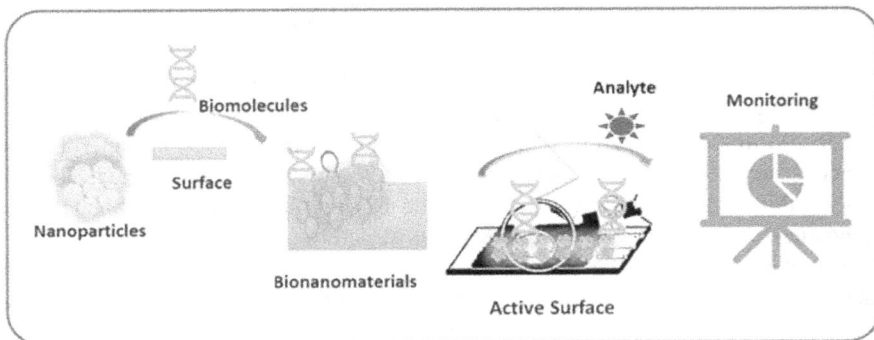

FIGURE 7.2   Schematic representation of bionanomaterials for analytes (food and cosmetic) monitoring.

aggregation, enhancing stability. Therefore, it favors the prompt monitoring of toxicants from food and cosmetic samples. Hence, resources can be designed using nanobiotechnology to improve global nutrition and cosmetic safety [24]. In recent years, bionanomaterials have been promising candidates for developing novel tools in food and cosmetic applications. Their structural properties and morphologies, such as small-size particles, and one-dimensional (nanotube, nanowire) and three-dimensional nanoflowers have exhibited advanced features compared with bulk materials.

7.5 MONITORING FOR ANALYTES IN FOOD AND COSMETIC SAMPLES

Bionanotechnology's advanced platform with biological molecules into nanoartifacts for food and cosmetic quality monitoring is gaining increased interest. The acute monitoring of samples helps ensure product safety and shelf life. Bionanomaterials' advanced and distinct properties are beneficial in determining quality, particularly for sprayed chemical products and added toxicants [25]. Combining biological principles with physical and chemical strategies generates useful functions with less toxicity and better biocompatibility. Currently, the safety and sustainability of food and daily-use products are critical issues that need more research [26]. Bionanotechnology has offered excellent contributions to food and cosmetics due to the unique properties of nanomaterials and biomolecules. In addition, it enables an intelligent detection system with enhanced sensitivity and specificity in real samples. Thus, product quality can be monitored and controlled more accurately. Furthermore, it can help in intelligent packaging systems and nanoencapsulation/target delivery with improved uptake by consumers [27].

7.5.1 FOOD SENSORS

Determining toxic analytes in foods and beverages such as yogurt, cheese, sour milk, beans, fruits, vegetables, cheese, processed meats, etc., is most important due to increasing food demand with a growing population. In addition, antioxidants and other chemical species have been used in food production to reduce oxidative processes in the human body. However, these chemical species can exhibit different adverse effects on human health [28,29]. For example, nanotechnology-based fibers with DNAzyme acted as a biosensor for ultra-trace detection of Pb^{2+} ions with LOD of 8.56 pM in the range 10^{-11} M to 10^{-6} M from real samples. The sensing platform was constructed by gold nanoparticle-Au with DNAzyme substrate strand. The Pb^{2+} ions exhibited catalytic activity, cleaved the DNAzyme-linked rA (ribonucleotide adenosine), and transformed the double-helix structure into a single strand [30].

A novel sensor can be developed with interest for the food industry using nanobiotechnology and nanomaterials. Developing functionalized nanomaterials as optical or electroactive platforms helps improve bio-sensing performances. Moreover, it provided a new platform with higher sensitivity, selectivity, and stability [31,32]. For example, the eggshell membranes designed by CdTe quantum dots (QDs) and

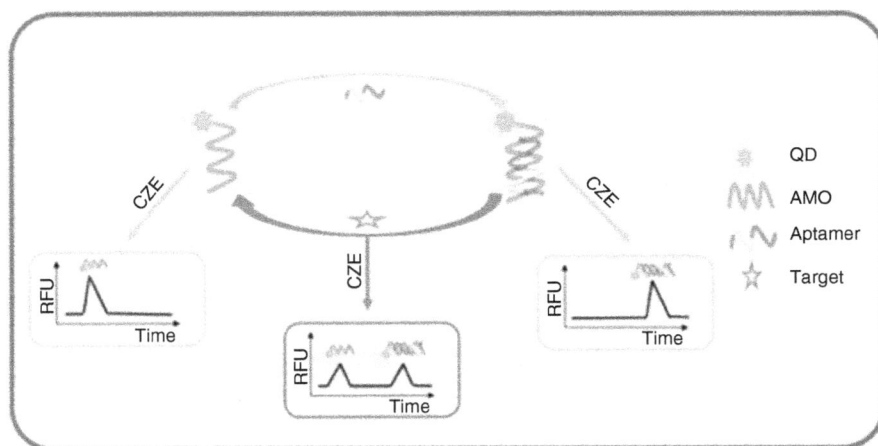

FIGURE 7.3 Bionanomaterial (quantum dot–DNA aptamer) strategy for ratiometric detection of organophosphorus pesticides. Reproduced with the permission from Ref. [34]. Copyright © 2015 Elsevier B.V. All rights reserved.

bi-enzyme-immobilized have been used to determine organophosphorus pesticide (OP) contaminants in food products. In a mechanistic study, OPs decreased the enzymatic activity that reduced hydrogen peroxide (H_2O_2) production. The designed material exhibited an LOD of 4.30×10^{-12} mol/L for paraoxon and 2.47×10^{-12} mol/L for parathion analytical parameters in the range of 1.0×10^{-11} to 1.0×10^{-6} mol/L [33]. In another example, DNA-functionalized CdTe/CdS QDs exhibited sensitive detection of OPs (phorate, profenofos, isocarbophos, and omethoate) with LODs of 0.20, 0.10, 0.17, and 0.23 μM, respectively, by capillary electrophoresis with laser-induced fluorescence. The QDs conjugated with the amino-modified oligonucleotide by amidation reaction that partially complemented the DNA aptamer of OPs, then incubated to form QD-AMO–aptamer duplex (Figure 7.3) [34]. Electrochemical technology with bionanomaterials has provided a tool as an immunosensor for low and sensitive detection of acrylamide food samples. Antibody- and AuNP-fabricated SnO_2-SiC nanomaterials possess high surface area and excellent electroconductivity. In addition, the composited material exhibited high specificity, reusability, and stability that demonstrated its potential application in acrylamide monitoring [35].

7.5.2 COSMETIC SENSORS

The cosmetic industry is a highly competitive market that uses different types of antioxidants and toxic chemicals in production (Figure 7.4). Therefore, it is necessary to use safe cosmetic products, but the question arises of how to monitor the product quality [36]. Bionanomaterial-based sensors have been used to monitor the toxicity level of cosmetic products. It has overcome conventional sensory methods' time-consuming, accuracy, and sensitivity-related limitations. Bionanomaterials are composite materials that exhibit enhanced optical and electrochemical signals. They

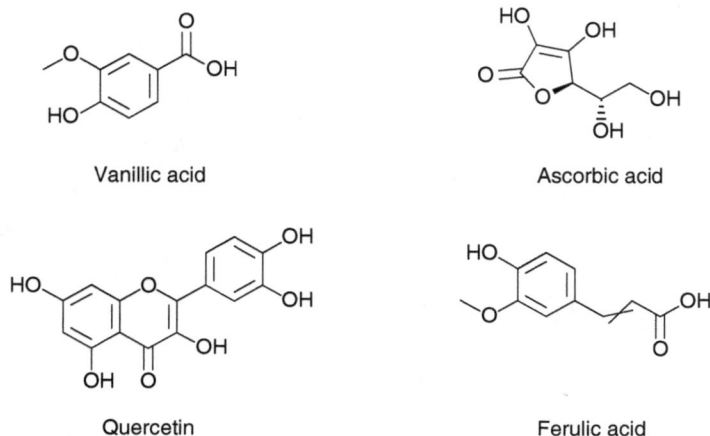

FIGURE 7.4   Common antioxidants used in cosmetic products.

can monitor the safety index products applied to various body parts (face, hair, lips, skin, etc.) [37].

Coumarin, a suspected carcinogen, is widely used in various cosmetic products. Fluorescence-based sensors that have been designed to determine coumarin rely on π–π stacking interaction as molecular recognition [38]. Also, an electrochemical detection strategy has been explored for sodium lauryl sulfate contaminants in cosmetics. The sensors have higher sensitivity, selectivity, and long-term stability which have attracted the attention of researchers to explore this area [39]. For example, Pereira and coworkers designed a hemoglobin-functionalized electrochemical biosensor for detecting H_2O_2 through oxidation of the iron ion. The biomolecule hemoglobin enhanced the detection efficiency and observed an LOD 4.0×10^{-6} mol/L, with the concentration range from 4.9×10^{-6} to 3.9×10^{-4} mol/L demonstrating the application of the designed biosensor (Figure 7.5) [40]. Another example is paraffin/graphite-modified with *Ipomoea batatas* (L.) Lam. The tissue-based biosensor detected H_2O_2 from cosmetic creams. This enzyme, in the presence of H_2O_2 oxidized hydroquinone to p-quinone. The analytical performance of the designed platform exhibited a high recovery rate of 99.1–104.1% with an LOD of 8.1×10^{-6} M [41].

7.6   KEY FINDINGS

Bionanomaterials with metal nanomaterials and biomolecules have a variety of applications. It is due to their excellent properties and combined electronic and structural properties that they proved to be advanced sensitive and stable biosensors. In addition, different metal and biomolecule-based materials have other working mechanisms due to their electronic properties. Based on their activity, the applications for detecting analytes can be determined, for example, the high functional weak acidic materials showed interactions with weak basic analytes and electron deficient materials for electron-rich analytes. Moreover, biomolecules can respond to external

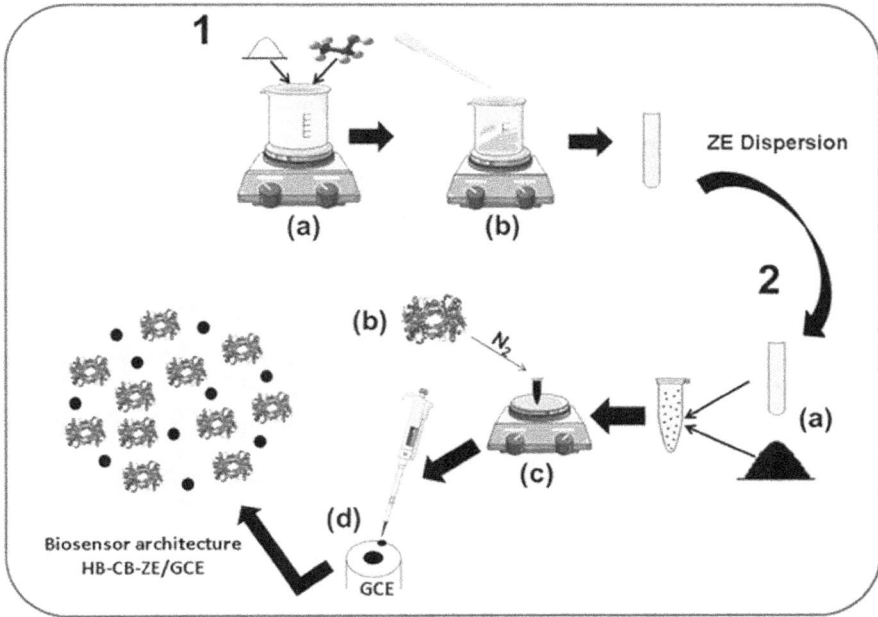

FIGURE 7.5 Hemoglobin-based electrochemical biosensors to detect H_2O_2 in cosmetic products. Reproduced with the permission from Ref. [40] © 2018 Elsevier B.V. All rights reserved.

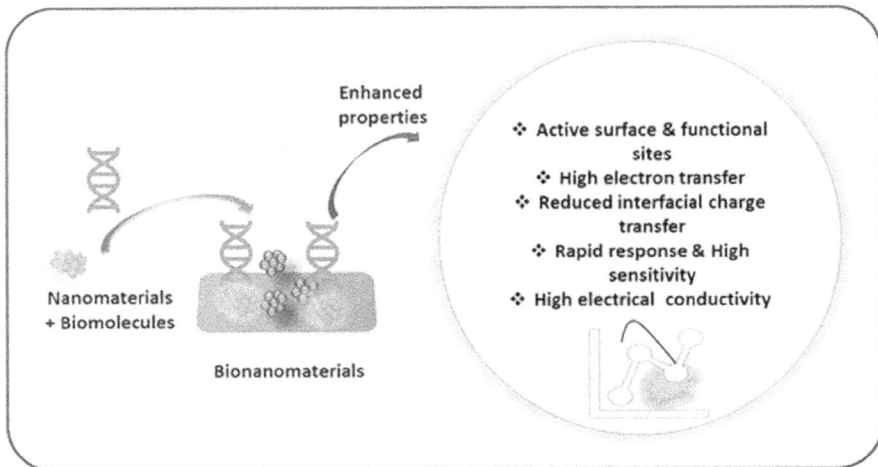

FIGURE 7.6 Bionanomaterials with enhanced properties.

stimuli that could help convert output signals. The fabrication of nanomaterials with biomolecules generated a new class of materials with enhanced properties (Figure 7.6). Hence, combining biomolecules with nanomaterials enhanced the biosensing activity of materials from food and cosmetic products that have significant applications in intelligent food monitoring [42].

7.7 CONCLUSION AND OUTLOOK

Designing new bionanomaterials for food and cosmetic products monitoring has enabled researchers to improve quality and safety. These basic nanotechnologies have been explored recently, and bionanotechnology has crucially emerged as an advanced tool for monitoring analytes. Bionanomaterials based on luminescence, electrochemical, and colorimetric analysis have been introduced for ultra-low-trace detection of hazardous chemicals from samples.

Moreover, the biomolecules with nanomaterials provided fast and quick devices for food and cosmetic monitoring. The efficiency and sensitivity of biomolecule-laced nanomaterials compared to traditional methods are promising detection tools. In conclusion, new bionanomaterials for food and cosmetic monitoring are advanced materials in new forms for developing current and future devices. This study will provide a significant route for academic and industrial purposes for food and cosmetic monitoring.

ACKNOWLEDGMENTS

The authors gratefully acknowledge the College of Ocean Food and Biological Engineering, Jimei University, Xiamen, China, for providing a research platform.

REFERENCES

[1] Singh, K., Verma, V., Yadav, K., Sreekanth, V., Kumar, D., Bajaj, A., et al. (2014) Design, regioselective synthesis and cytotoxic evaluation of 2-aminoimidazole–quinoline hybrids against cancer and primary endothelial cells. *European Journal of Medicinal Chemistry* 87: 150–158.

[2] Ouedraogo, G., Alexander-White, C., Bury, D., Clewell, H.J., Cronin, M., Cull, T., et al. (2022) Read-across and new approach methodologies applied in a 10-step framework for cosmetics safety assessment – A case study with parabens. *Regulatory Toxicology and Pharmacology* 132: 105161.

[3] Barabadi, H., Ovais, M., Shinwari, Z.K., Saravanan, M. (2017) Anti-cancer green bionanomaterials: present status and future prospects. *Green Chemistry Letters and Reviews* 10: 285–314.

[4] Rothen-Rutishauser, B., Bogdanovich, M., Harter, R., Milosevic, A., Petri-Fink, A. (2021) Use of nanoparticles in food industry: current legislation, health risk discussions and public perception with a focus on Switzerland. *Toxicological & Environmental Chemistry* 103: 420–434.

[5] Jain, B., Singh, A.K., Bin, M.A., Susan, H. (2021) *Plastics and e-Waste, a Threat to Water Systems*, pp. 119–130. Springer.

[6] Yilmaz, B., Keskinates, M., Bayrakci, M. (2021) Novel integrated sensing system of calixarene and rhodamine molecules for selective colorimetric and fluorometric detection of Hg^{2+} ions in living cells. *Spectrochimica Acta Part A: Molecular and Biomolecular Spectroscopy* 245: 118904.

[7] Singh, G., Priyanka, A. Singh, P. Satija, Sushma, P., et al. (2021) Schiff base-functionalized silatrane-based receptor as a potential chemo-sensor for the detection of Al^{3+} ions. *New Journal of Chemistry* 45: 7850–7859.

[8] Hashmi, A., Nayak, V., Singh, K.R., Jain, B., Baid, M., Alexis, F., et al. (2022) Potentialities of graphene and its allied derivatives to combat against SARS-CoV-2 infection. *Materials Today Advances* 13: 100208.

[9] Jain, B., Singh, A.K., Hashmi, A., Susan, M.A.B.H., Lellouche, J.P. (2022) Surfactant-assisted cerium oxide and its catalytic activity towards Fenton process for non-degradable dye. *Advanced Composites and Hybrid Materials* 3: 430–441.

[10] Mohan, B., Kumar, S., Ma, S., You, H., Ren, P. (2021) Mechanistic insight into charge and energy transfers of luminescent metal–organic frameworks based sensors for toxic chemicals. *Advanced Sustainable Systems* 2000293.

[11] Jain, B., Hashmi, A., Sanwaria, S., Singh, A.K., Susan, M.A.B.H, Singh, A. (2020) Zinc oxide nanoparticle incorporated on graphene oxide: an efficient and stable photocatalyst for water treatment through the Fenton process. *Advanced Composites and Hybrid Materials* 3: 231–242.

[12] Mohan, B., Kumar, S., Sharma, H.K. (2020) Synthesis and characterizations of flexible furfural based molecular receptor for selective recognition of Dy(III) ions. *Polyhedron* 183: 114537.

[13] Jain, B., Gade, J.V., Hadap, A., Ali, H., Katubi, K.M., Sasikumar, B., et al. (2022) A facile synthesis and properties of graphene oxide-titanium dioxide-iron oxide as Fenton catalyst. *Adsorption Science & Technology* https://doi.org/10.1155/2022/2598536

[14] Aggarwal, R., Singh, G., Kumar, S. (2021) Molecular iodine mediated transition-metal-free oxidative dehydrogenation of 4,7-dihydropyrazolo[3,4-b]pyridines. *An International Journal for Rapid Communication of Synthetic Organic Chemistry* 51: 3601–3609.

[15] Aggarwal, R., Kumar, S., Mittal, A., Sadana, R., Dutt, V. (2020) Synthesis, characterization, in vitro DNA photocleavage and cytotoxicity studies of 4-arylazo-1-phenyl-3-(2-thienyl)-5-hydroxy-5-trifluoromethylpyrazolines and regioisomeric 4-arylazo-1-phenyl-5(3)-(2-thienyl)-3(5)-trifluoromethylpyrazoles. *Journal of Fluorine Chemistry* 236: 109573.

[16] Naghdi, T., Golmohammadi, H., Yousefi, H., Hosseinifard, M., Kostiv, U., Horák, D., et al. (2020) Chitin nanofiber paper toward optical (bio)sensing applications. *ACS Applied Materials Interfaces* 12: 15538–15552.

[17] Potyrailo, R., Naik, R.R. (2013) Bionanomaterials and bioinspired nanostructures for selective vapor sensing. http://dx.doi.org/10.1146/annurev-matsci-071312-121710 43: 307–334.

[18] Dhanjal, D.S., Mehra, P., Bhardwaj, S., Singh, R., Sharma, P., Nepovimova, E. et al. (2022) Mycology–nanotechnology interface: Applications in medicine and cosmetology. *International Journal of Nanomedicine* 17: 2505–2533.

[19] Huang, X., Zhu, Y., Kianfar, E. (2021) Nano biosensors: Properties, applications and electrochemical techniques. *Journal of Materials Research and Technology* 12: 1649–1672.

[20] Kaur, R., Khullar, P., Gupta, A., Bakshi, M.S. (2021) Extraction of bionanomaterials from the aqueous bulk by using surface active and water-soluble magnetic nanoparticles. *Langmuir* 37: 14558–14570.

[21] Kaur, M., Arshad, M., Ullah, A. (2018) In-situ nanoreinforced green bionanomaterials from natural keratin and montmorillonite (MMT)/cellulose nanocrystals (CNC). *ACS Sustainable Chemistry & Engineering* 6: 1977–1987.

[22] Parshad, B., Yadav, P., Kerkhoff, Y., Mittal, A., Achazi, K., Haag, R., et al. (2019) Dendrimer-based micelles as cyto-compatible nanocarriers. *New Journal of Chemistry* 43: 11984–11993.

[23] Yu, L., Si, P., Bauman, L., Zhao, B. (2020) Synergetic combination of interfacial engineering and shape-changing modulation for biomimetic soft robotic devices. *Langmuir* 36: 3279–3291.

[24] Lugani, Y., Sooch, B.S., Singh, P. Kumar, S. (2021) Nanobiotechnology applications in food sector and future innovations. *Microbial Biotechnology: Food and Health* 197–225.

[25] Pan, Y., Zhou, S., Liu, C., Ma, X., Xing, J., Parshad, B., et al. (2022) Dendritic polyglycerol-conjugated gold nanostars for metabolism inhibition and targeted photothermal therapy in breast cancer stem cells. *Advanced Healthcare Materials* 11: 2102272.

[26] Parshad, B., Kumari, M., Khatri, V., Rajeshwari, R., Pan, Y., Sharma, A.K., et al. (2021) Enzymatic synthesis of glycerol, azido-glycerol and azido-triglycerol based amphiphilic copolymers and their relevance as nanocarriers: A review. *European Polymer Journal* 158: 110690.

[27] Sonali, M.I., Sonali J., Subhashree S., Senthil Kumar, P., Veena Gayathri, K. (2022) New analytical strategies amplified with carbon-based nanomaterial for sensing food pollutants. *Chemosphere* 295: 133847.

[28] Petrucci, R., Bortolami, M., Di Matteo, P., Curulli, A. (2022) Gold nanomaterials-based electrochemical sensors and biosensors for phenolic antioxidants detection: Recent advances. *Nanomaterials* 12: 959.

[29] Zhang, H., Wei, X., Chan-Park, M.B., Wang, M. (2022) Colorimetric sensors based on multifunctional polymers for highly sensitive detection of food spoilage. *ACS Food Science & Technology* 2: 703–711.

[30] Wang, F., Zhang, Y., Lu, M., Du, Y., Chen, M., Meng, S., et al. (2021) Near-infrared band Gold nanoparticles-Au film "hot spot" model based label-free ultratrace lead (II) ions detection via fiber SPR DNAzyme biosensor. *Sensors Actuators B Chemistry* 337: 129816.

[31] Xu, P., Ghosh, S., Gul, A.R., Bhamore, J.R., Park, J.P., Park, T.J. (2021) Screening of specific binding peptides using phage-display techniques and their biosensing applications. *TrAC Trends in Analytical Chemistry* 137: 116229.

[32] Zhou, R., Zhao, L., Wang, Y., Hameed, S., Ping, J., Xie, L., et al. (2020) Recent advances in food-derived nanomaterials applied to biosensing. *TrAC Trends in Analytical Chemistry* 127: 115884.

[33] Xue, G., Yue, Z., Bing, Z., Yiwei, T., Xiuying, L., Jianrong, L. (2016) Highly-sensitive organophosphorus pesticide biosensors based on CdTe quantum dots and bi-enzyme immobilized eggshell membranes. *Analyst* 141: 1105–1111.

[34] Tang, T., Deng, J., Zhang, M., Shi, G., Zhou, T. (2016) Quantum dot-DNA aptamer conjugates coupled with capillary electrophoresis: A universal strategy for ratiometric detection of organophosphorus pesticides. *Talanta* 146: 55–61.

[35] Wu, M.F., Wang, Y., Li, S., Dong, X.X., Yang, J.Y., Shen, Y.D., et al. (2019) Ultrasensitive immunosensor for acrylamide based on chitosan/SnO_2-SiC hollow sphere nanochains/gold nanomaterial as signal amplification. *Analytica Chimica Acta* 1049: 188–195.

[36] Vieira, G.S., Lavarde, M., Fréville, V., Rocha-Filho, P.A., Pensé-Lhéritier, A.M. (2020) Combining sensory and texturometer parameters to characterize different type of cosmetic ingredients. *International Journal of Cosmetic Science* 42: 156–166.

[37] Fairhurst, D., Dukhin, A.S., Klein, K. (2004) *Stability and Structure Characterization of Cosmetic Formulations Using Acoustic Attenuation Spectroscopy*, pp. 249–269. ACS.

[38] Yang, H., Chen, X., Wu, J., Wang, R., Yang, H. (2019) A novel up-conversion fluorescence resonance energy transfer sensor for the high sensitivity detection of coumarin in cosmetics. *Sensors Actuators B Chemistry* 290: 656–665.

[39] Motia, S., Tudor, I.A., Madalina Popescu, L., Piticescu, R.M., Bouchikhi, B., El Bari, N. (2018) Development of a novel electrochemical sensor based on electropolymerized molecularly imprinted polymer for selective detection of sodium lauryl sulfate in environmental waters and cosmetic products. *Journal of Electroanalytical Chemistry* 823: 553–562.

[40] dos Santos Pereira, T., Mauruto de Oliveira, G.C., Santos, F.A., Raymundo-Pereira, P.A., Oliveira, O.N., Janegitz, B.C. (2019) Use of zein microspheres to anchor carbon black and hemoglobin in electrochemical biosensors to detect hydrogen peroxide in cosmetic products, food and biological fluids. *Talanta* 194: 737–744.

[41] Cruz Vieira, I., Fatibello-Filho, O. (2000) Biosensor based on paraffin/graphite modified with sweet potato tissue for the determination of hydroquinone in cosmetic cream in organic phase. *Talanta* 52: 681–689.

[42] Yang, J., Shen, M., Luo, Y., Wu, T., Chen, X., Wang, Y., et al. (2021) Advanced applications of chitosan-based hydrogels: From biosensors to intelligent food packaging system. *Trends in Food Science & Technology* 110: 822–832.

8 Bionanotechnology in Forensic Science

Pradip Hirapure and Arti Shanaware

8.1 INTRODUCTION

A variety of physical, chemical, and biological sciences are combined in nanotechnology to research phenomena at the nanometer scale (1 nm = 1 billionth of a meter) [1]. A few years ago, Nanotechnology was the subject of fanciful studies in secret labs and fantasy novels. Among the technologies currently in use, nanotechnology is one of the most promising yet debatable developing technologies [2]. Given how easy it is to collect, examine, and find complicated nanoparticles and buried pieces of evidence from the scene of a crime, nanotechnology's introduction to forensic science has only recently begun to show itself as a game changer. The speed of evidence analysis is accelerated by applied nanotechnology in genetics, medicine, and analytical chemistry, and real-time evidence analysis has been used in forensic sciences [3,4]. Today, nanotechnology is making a significant contribution to science. According to one definition, it is "the study, design, development, synthesis, manipulation, and application of functional materials, devices, and systems through control of matter at the nanoscale scale" [5].

Television shows like Crime Scene Investigation (CSI) and National Crime Investigation Service (NCIS) have demonstrated how technically complex and professional the field of forensics is. The development of analytical methods has been the main force behind the advancements in forensic technology. The most well-known area is DNA analysis, while advancements in the sensitivity of older, more traditional analytical methods have also contributed [6]. Police forces and other law enforcement agencies are starting to assess and deploy these nanotechnology-based tactics. The ability of the new methodologies to give either increased performance over already available materials or the ability to gather information from a crime scene that would not otherwise be achievable is made possible by the nanoscale characteristics of the materials.

There are various forensic science fields to look into, including latent fingerprint development, the detection of illegal drugs, the measurement of alcohol in drunk drivers, the detection of explosives, nerve gas, saliva, the identification of inorganic pigments in hit-and-run accidents, and many other applications where nanotechnology plays a significant role in ensuring accurate results and time-bound

DOI: 10.1201/9781003362258-8

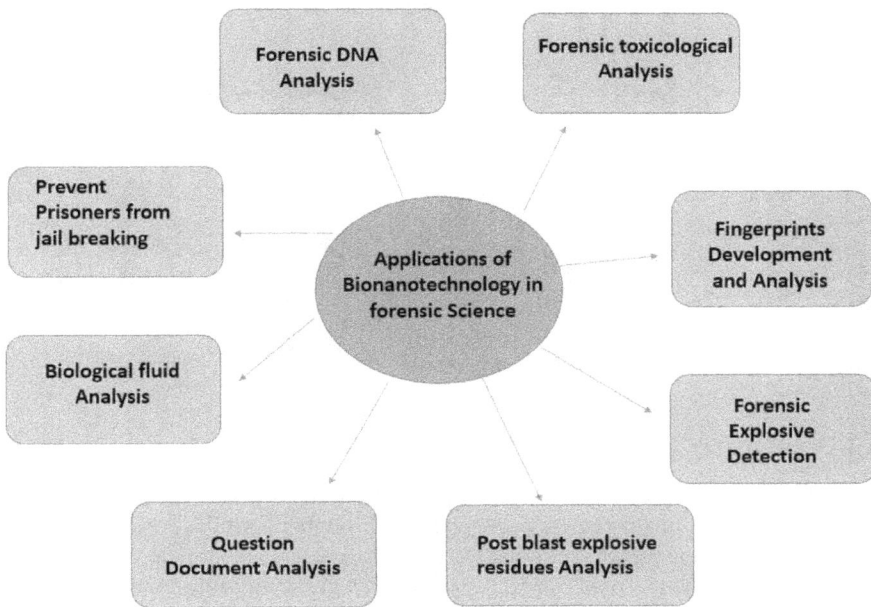

FIGURE 8.1 Applications of nanotechnology in forensic science.

investigation [7,8]. For accurate results and timely investigations, nanotechnology is also essential in cases involving the identification of illegal drugs, the identification of inorganic pigment in hit-and-run accidents, the development of latent fingerprints, the detection of nerve gases, the measurement of alcohol in cases of drink-driving, and other situations of this nature. Nanotechnology is free of human biases, which has boosted the speed and precision of research. Therefore, a brief assessment of nanotechnology in relation to its use in the forensic sector is made and illustrated in Figure 8.1 [9].

8.2   NANOTECHNOLOGY IN FORENSIC DNA ANALYSIS

The post-PCR (polymerase chain reaction) quantification method is currently the most widely used forensic application of microfluidic systems. The commercially available Agilent 2100 bioanalyzer injects and measures nanoliter quantities of 12 double-stranded DNA samples in half an hour using a variety of different channels. These systems, which have run times of under 2 minutes per sample, are currently being employed in various forensic laboratories to perform post-PCR quantification of mitochondrial DNA. The possibility of such devices being employed at the crime scene is frequently raised because of their modest size. In addition, the technology will advance due to the simplicity of the quick, disposable devices that need little upkeep [10]. For the extraction of copper, magnetic nanoparticles, particularly magnetic nanoparticles based on silica, are helpful [11,12].

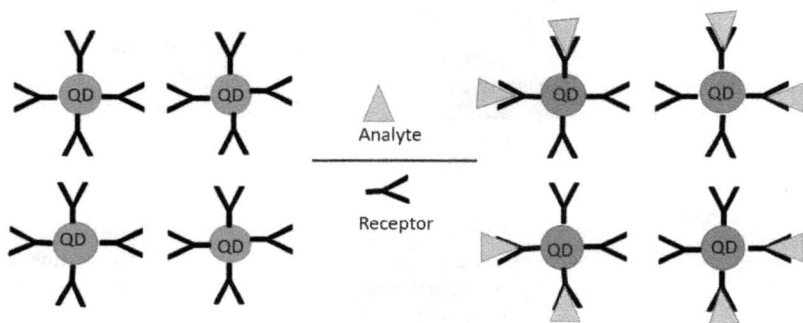

FIGURE 8.2   Quantum dots as nanosensors for the detection of toxic compounds.

8.3 NANOTECHNOLOGY IN FORENSIC TOXICOLOGY ANALYSIS

The investigation of various harmful components from forensic evidence such as hair, saliva, blood, and human bones is greatly helped through the application of gold, silver, and titanium dioxide nanotechnology [13,14]. Different toxicological substances are screened using nanosensors. The nanosensor is a significant replacement for the on-field test, and can be employed in the process of instantaneous spot testing. It is a low-cost, active, stable, and time-specific method of drug testing for forensic purposes. The usefulness of nanosensors for toxicological analysis was clarified by the practical application of forensic nanotechnology to actual specimens [14]. Due to their high sensitivity, selectivity, short testing times, and low use of chemical reagents, sensor-based systems for chemical analysis have achieved significant advancements. Great advances have been made in sensor-based methods for chemical analysis owing to their high sensitivity, selectivity, shorter testing time, and the minimal usage of chemical reagents. The excellent optical properties of quantum dots (QDs) have made them widely investigated for a variety of scientific applications where light plays a key role, as shown in Figure 8.2. The use of QDs as photoluminescent nanosensors for the detection of chemicals and biomolecules has received increasing attention in recent years. To open a fresh viewpoint on the role of sensors in forensic toxicology, a literature review on the uses of QDs as chemosensors for the detection of gaseous, anionic, phenolic, metallic, drug-overdose, and pesticide poison has been conducted.

8.4 NANOTECHNOLOGY IN GUNSHOT RESIDUE ANALYSIS

The analysis of gunshot residue (GSR) is a critical step in forensic studies of shooting and related criminal cases. However, the current techniques used for GSR analysis are not complete. Detailed information regarding the elemental and crystallographic signatures of GSR are missing. Moreover, the analysis requires a substantial amount of sample, which can be difficult to obtain and frequently might be contaminated. Electron microscopic studies of metallic nanoparticles (10–100 nm in diameter) obtained from GSR at different target distances from a Winchester Super-X 9 mm luger have been analyzed in detail. Perfectly spherical (diameter ~ 10 nm) and very crystalline Pb and Sb nanoparticles were observed. Theoretical studies explaining the formation of these nanoparticles is reported. The non-equilibrium thermodynamic processes leading to the synthesis of nanoparticles was observed to be very similar to the artificial chemical synthesis methods (e. g. CVD, laser ablation etc.) [15–19].

8.5 NANOTECHNOLOGY IN POST-BLAST EXPLOSIVE RESIDUES ANALYSIS

The largest social enemy on the planet right now is terrorism. Its rapid expansion is due to the ease with which explosive-based weapons and instruments may be produced and deployed, and the catastrophic destruction that results [20]. Due to a variety of problems, including a dearth of unexploded explosives, contaminated samples, and diverse sample collection methods, the detection of tiny levels of explosives is a difficult undertaking. In the event of a bomb blast, fragmented explosive leftovers may be dispersed away from the blast site, while an intact portion of the explosive may remain at the scene of the crime. Investigators can employ nanotechnology to identify unfragmented/trace amounts of fragmented explosives from a crime scene during the crime scene investigation. In the majority of bomb blast incidents, it is challenging for the investigator to identify unfragmented explosives. As a result, they are unable to present the court with sufficient evidence to support the connection between the accused and the crime scene, which is also insufficient to support a conviction [21]. Here, nanotechnology-based explosive case analyses may be helpful. Investigators are able to identify minuscule gun powder particles on the shooter's hand thanks to nanotechnology [22]. Investigators are also unable to locate even a single particle of the gunshot residue after the crime due to the use of fruitless technologies for gunshot residue detection. Gunshot residue can be found using high-resolution scanning electron microscope (SEM) imaging, while identifying gunshot residue with the use of X-ray spectrometry may be beneficial. According to a recent study, gunshot residue analysis is a crucial stage in shooting cases that could contribute to crime prevention by providing sufficient corroborating evidence [23,24]. These procedures take time, call for special equipment, and increase the likelihood of error. Additionally, explosives fragments and traces of explosive residue particles left at the blast site have been found and collected with the use of nanotechnology. Additionally, it can be used to find non-fragmented explosives [2,15]. According to Pandya et al. (2012), curcumin nanoparticles derived from turmeric are a very sensitive fluorescent dye that can be used for the trinitrotoluene (TNT) detection up to the 1 nM level in an aqueous solution (Figure 8.3). Chu et al. (2015) proposed a novel sensing technique for a label-free and selective detection of 2,4,6-trinitrotoluene (TNT) from 10^{-12} to 10^{-4} M based on amine-terminated nanoparticles. These studies open up a new prospect for rapid, easy, and reliable detection of TNT from environmental samples and also at crime scenes by measuring the fluorometric intensity.

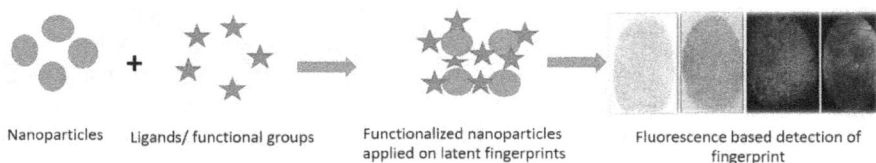

| Nanoparticles | Ligands/ functional groups | Functionalized nanoparticles applied on latent fingerprints | Fluorescence based detection of fingerprint |

FIGURE 8.3   Latent fingerprint developed on various nonporous/semiporous surfaces using functionalized nanoparticles.

8.6   NANOTECHNOLOGY IN FORENSIC EXPLOSIVE DETECTION

Nanomaterials have the active potential to create explosives detection sensors. Advanced nanosensor concept devices, such as electronic noses, nanotubes, and nano-mechanical devices, are accustomed to detecting conventional bombs, plastic explosives, and grenades in order to trace explosives. Typically, an electronic nose has a chemical detecting system like an artificial neural network [24–27]. An alteration in the characteristics of the nanoparticles can be achieved by targeting nanoparticle–ligand complexes to particular analytes. Due to their small size, nanoparticles have electromagnetic properties that can be changed by analyte binding and used as a transducer in a chemical-sensing system for explosive analytes. The surface plasmon resonance (SPR) band and fluorescence of colloidal gold nanoparticles (AuNPs) are the first and the second is colloidal semiconductor fluorescence (quantum dots [QDs]). Picomolar or lower concentrations of explosive analytes may be detected by sensors based on these nanoparticle characteristics, and they can function for both solution-phase and gas-phase detection [1–3]. It is also simpler to construct full detection systems from standard components because the signal produced by the nanoparticle transducer is measured using normal scientific apparatus.

8.7   NANOTECHNOLOGY IN FINGERPRINT DEVELOPMENT AND ANALYSIS

Since ancient times, fingerprints have been used as unique evidences which were later also used on Babylonian clay tablets for business transactions [1]. Ideally, the fingerprint powder will fix to the residues left by the finger and give rise to the distinctive patterns that help to identify an individual as a fingerprint, but not stick to the background or other surfaces. Latent fingerprints are commonly developed by various colored materials such as carbon black on a white/light background and aluminum flakes on a black/dark background. The drawback of these materials is their adherence power because they do not only help to decipher the latent fingerprint but also stick to the background of the fingerprint. Since they adhere to both the print and its background, it is difficult to obtain a clear image of the fingerprint, so identification becomes a major issue. In this condition, engineered or synthesized nanoparticles are being used to overcome this problem. In forensic science, nanotechnology can be helpful in the collection and analysis of evidence found at the crime scene. Forensic experts gather an extensive range of physical evidence from the scene of crime that could be helpful in correlating a suspect with the crime. These physical evidences not only link an individual/group of individuals with the crime but could help to prove them innocent. In the normal procedure, after collecting the evidence, it is packed appropriately and sent to the laboratory for further analysis. These procedures are time-consuming and the chances of error are increased. In addition, a study revealed that the micro-X-ray fluorescence method can be used to visualize latent fingerprints [28]. In this effective method, the latent fingerprint becomes visualized by ascertaining inorganic constituents that are present in the prints. This method contrasts with the chemical enhancement methods that have been applied to visualize latent fingerprints. This micro-X-ray fluorescence method has several advantages over conventional chemical methods for detecting latent fingerprints. Micro-X-ray fluorescence is a non-destructive method that does not affect the analysis or stability of the inorganic residues during image formation of fingerprints. During analysis of the fingerprint, results have revealed that some of the residues contain silicon, aluminum, and calcium [29]. This method still requires instruments and trainers to conduct such analysis. However, the latest nanotechnology-based techniques can help to analyze the evidence on the spot at the scene of the crime, which not only saves the time of analysis but also reduces the chances of error. Nowadays, in the forensic investigation process, different types of nanopowders have been applied to reveal the latent fingerprints on various surfaces [30]. In this respect, photoluminescent CdS semiconductor nanocrystals capped with dioctysulfo-succinate have been used to improve the detection of fingerprints [31]. Recently, a group of researchers synthesized novel ZnO-SiO$_2$ nanopowder using the conventional heating method [32]. The fingerprints were developed by powder dusting and small particle reagent (SPR) methods. This method has been effectively applied on various dry (semi-porous and non-porous) and wet (non-porous) surfaces to visualize latent fingerprints, as shown in Figure 8.3. The results exhibited that ZnO–SiO$_2$ nanopowder has excellent potential to envisage finger ridge detail at a higher level that shows superior discernibility than other commercially available white powders.

The most advanced applications are those associated with fingerprints. Substituting materials to develop fingerprints (such as lampblack, aluminum flake, and gentian violet) with much smaller nanoparticles increases by orders of magnitude the sensitivity of the forensic search. This makes it easier to detect old or faint fingerprints, and those left on difficult surfaces, such as adhesive or textured ones. Nanoparticles bind with the fingerprint pattern and make it visible [33]. A typical strategy is engineering fluorescent nanoparticles to make the development easier. However, the improvement is not only in sensitivity. Nanoparticles can reveal information held by fingerprints that is currently inaccessible. In particular, they can be engineered to bring antibodies that bind with metabolites contained in the sweat of the fingerprint [34]. These metabolites are breakdown products of the consumption of cocaine or nicotine. When the antibodies recognize the metabolite, they cause a color change [34]. This allows for "life-style intelligence"; that is, deducing elements of the lifestyle of people from the traces left on the things they touch. The company Intelligent Fingerprinting Limited, a spin-off of the University of East Anglia, has applied this idea into a transportable device which can quickly detect drug use through fingerprints [35].

8.8   NANOTECHNOLOGY IN QUESTIONABLE DOCUMENT ANALYSIS

If a person is discovered having completed suicide in a room with a written note in the body's vicinity, and the note has overlapping writing raising the suspicion of it having the handwriting of two or more different individuals, under such circumstances, nanotechnology can aid the investigating officer through a nanotool called the "atomic force microscope" (AFM). This tool helps forensic scientists to study the surface of paper at the nanoscale [36]. It provides information on the pen, ink, pressure/intensity exerted while writing, ink crossing etc., helping the investigator to determine whether the document is a forgery or if it was written by one or more persons.

8.9   NANOTECHNOLOGY IN BIOLOGICAL FLUID ANALYSIS

The atomic force microscope (AFM) also aids the investigating officer to investigate bodily offences, by revealing the age of a blood sample. With time blood becomes thicker and stiffer. AFM, by measuring the viscosity or dryness, can disclose the date of the sample. AFM additionally helps investigators in revealing the substances present within urine. When urine is mixed with gold nanoparticles and illuminated by a laser, the signal emitted announces the chemicals or materials, such as drugs, present within the urine. As an example, this can help the investigator to know if someone sexually assaulted an individual after administering a drink spiked with a so-called rape drug, even several days following the incident [36].

8.10   NANOTECHNOLOGY TO PREVENT PRISONERS FROM ESCAPING JAIL

Use of bar codes and trackers has become common. Trackers can help track down items which go missing or are stolen. Nanotrackers are also being used to prevent prisoners from escaping jail and to monitor prisoners after their release. Prisoners injected with nanotrackers are easily trackable [37].

8.11   NANOTECHNOLOGY AND ITS MILITARY USAGE

Nanotechnology often acts as an enabling means that enhances applications in numerous ways. Defense applications for nanotechnology are numerous, ranging from sensing weapons of mass destruction (WMD), combatant protection kits (smart armor, active camouflage), and medical aid (infection control), to self-healing materials and nano-electronics. At the same time, nanotechnology is an evolving technology and there are many theoretical possibilities for applications in the military realm which demand more in-depth research and development [38].

TABLE 8.1

Applications of Nanotechnology in Forensic Science

S. no.	Nanotechnologies	Applications	References
1.	Metal nanoclusters of sensors, microchips embedded in polymer layers containing specific ligands. AuNPs in alkanethiols layer	Warfare agent detection: calorimetric change in color in presence of warfare agents	[40]
2.	Biosensors (antigen–antibody interaction mechanism) Magnetic nanoparticles (NPs), AuNPs, Ln-doped NPs, quantum dots, Ag nanorods, nanowires, carbon nanotubes	Biological warfare agent detection, e.g., variola virus, bacterial species, *Brucella* spp., BoNT, etc.	[41,42]
3.	Carbon NTs, graphene, nanodiamonds, fluorescence sensors, nanoplatforms	Chemical warfare agent detection	[43]
4.	Lab-on-a-chip, AuNPs, triarylcarbinol functionalized AuNPs	Calorimetric detection of warfare and nerve agents	[44–49]
5.	Ag-NPs and tetraphenylene probe and silica NPs	H_2O_2 detection	[50]
6.	Chromophore: chromosensors AuNPs/QDs, nanoprobes	Ricin detection	[51]
7.	Turmeric extracted curcumin NPs: fluorescent probe Amine-terminated NPs	TNT detection up to 1 Nm in aqueous solution	[52]
8.	AuNPs and QD	DNT, TNT, RDX, PETN, tetryl detection	[53,54]
9.	Cellulose nanocrystals	Nitrophenolic explosives detection	[55]
10.	Electronic noses, sensors, nanowires, nanochips, nanotubes, nano mechanical devices	Explosives detection	[54,56]
11.	Tetraphenylethene probe-based fluorescent silica NPs	Nitramine explosive detection	[57]
12.	Luminescent sensors, carbon dots	Nitroaromatic explosive detection	[58,59]
13.	Cysteine-modified AuNP-SERS probe (free surface-enhanced Raman spectroscopy)	TNT detection	[60]
14.	CdS semiconductor nanocrystal capped dioctysulfo-succinate, ZnO-SiO$_2$ nanopowder SPR method	Development of latent fingerprint	[52]
15.	ZnO (20 nm), CdSe/ZnS (10 nm) nano powder in conjugation with SALDI-TOF- MS	Generate UV fluorescent fingerprints on development	[40,54,57,61–64]

(Continued)

TABLE 8.1 (Continued)
Applications of Nanotechnology in Forensic Science

S. no.	Nanotechnologies	Applications	References
16.	Gold NPs capped with antibody against cotinine (metabolite of nicotine)	Along with fingerprint development, deter mines if the person is a smoker or not.	[40,65]
	Silica stable NPs	Age and ethnicity of fingerprint donor	
17.	TiO$_2$-NPs (show fluorescence when conjugated with dyes)	Develop fingerprints on porous and non-porous surfaces	[63,66]
18.	Molybdenum disulfide NPs	Fingerprint analysis: forms gray deposits on reaction with fatty acids	[40,54]
19.	Gold NPs capped with citrate ions/AuNPs with Ag-PDA	Silver metallic prints obtained on analysis of latent fingerprints	[40,57,61]
20.	Silver NPs (1–200 nm size), aluminum oxide NPs (30–60 nm size)	Latent FP analysis	[54,62,67,68]
		Al-NPs give good contrast in wet conditions	
21.	Eu^{3+} doped Al$_2$O$_3$ nanocrystalline powder (36.8 nm average)	Latent fingerprint development	[54,62,63,68]
	Fluorescent starch-based carbon NPs (10–210 nm)	Enhanced contrast on non-porous surface	
22.	Poly(styrene-alt maleic anhydride)-b-polystyrene (SMA-b-PS) functionalized gold NPs	Latent fingerprint development	[64]
	Quantum dots		[54,57]
23.	Citrate stabilized AuNPs with smart phone camera	Codeine sulfate detection	[52]
24.	AuNPs in presence of melamine	Clonazepam detection	[52,54]
25.	Gold and silver NPs	Cocaine detection in fingerprints	[69]
26.	Nanoprobe-based detection	Detection of morphine, codeine, methamphetamine, clonazepam	[70]
	Nanochips-based detection	Detection of gamma hydroxybutyrate, benzoylecgonine, cocaine	
27.	Mix and detect: Thioflavin with anticocaine aptamer MNS 4.1	Cocaine detection (fluorescence)	[71]

8.12 FUTURE AREAS FOR NANOTECHNOLOGY IN FORENSIC SCIENCE

The integration of these scientific areas provides advantages in the development of nanotechnology in various areas of forensic science, health sciences, and automotive engineering (Table 8.1) by combining the most advanced chemical and physical technologies with the needs of modern applications of biomedical and forensic research. The growing demand for nanotechnology today has enabled scientists and analysts to gain efficient strategic objectives and sound skills in the field of nanotechnology [39].

8.13 CONCLUSION

This chapter describes how nanotechnologies may be important for solving current forensic investigation problems. The field of forensic science gains greatly from nanotechnology. It is particularly beneficial in unearthing concealed evidence. Nanotechnology aids forensic scientists in establishing the veracity of their investigation's findings in legal proceedings. It is critical to recognize that science is advancing quickly and continuously. As new technologies are discovered, it is essential for the forensic investigator to keep up with changes in the industry. The significance of nanotechnology in the field of forensic science and its applications during the various stages of forensic investigation are clearly explained in this chapter.

REFERENCES

1. Pitkethly, M. (2010) Nanotechnology and forensics. *Materials Today* 12(6). Retrieved from: www.materialstoday.com/nanomaterials/articles/s1369702109701671/
2. Pandya, A., Shukla, R.K. (2018). New perspective of nanotechnology: role in preventive forensic. *Egyptian Journal of Forensic Sciences* 8(1): 1–11.
3. Chauhan, V., Singh, V., Tiwari, A. (2017). Applications of nanotechnology in forensic investigation. *International Journal of Life-Sciences Scientific Research* 3(3): 1047–1051.
4. Ganesh, E.N. (2016) Application of nanotechnology in forensic science. International *Journal of Printing, Packaging & Allied Sciences* 4: 5.
5. Ganesh, E.N. (2016). Application of nanotechnology in forensic science. *Environment* 1(1).
6. Romeika, J.M., Yan, F. (2013) Recent advances in forensic DNA analysis. *Journal of Forensic Research* 12(001).
7. Dhawan, A., Sharma, V., Parmar, D. (2009) Nanomaterials: a challenge for toxicologists. *Nanotoxicology* 3(1): 1–9.
8. Sharma, V., Shukla, R.K., Saxena, N., Parmar, D., Das, M., Dhawan, A. (2009) DNA damaging potential of zinc oxide nanoparticles in human epidermal cells. *Toxicology Letters* 185(3): 211–218.
9. Hallikeri, V.R., Bai, M., Kumar, A.V. (2012) Nanotechnology—The future armour of forensics: A short review. *Journal of the Scientific Society* 39(1): 10.
10. McCord, B. (2006). Nanotechnology and its potential in forensic DNA analysis. *Profiles in DNA* 9(2): 7–9.
11. Yi, L., Huang, Y., Wu, T., Wu, J. (2013) A magnetic nanoparticles-based method for DNA extraction from the saliva of stroke patients. *Neural Regeneration Research* 8(32): 3036.

12. Yang, Y., Xie, B., Yan, J. (2014) Application of next-generation sequencing technology in forensic science. *Genomics, Proteomics & Bioinformatics* 12(5): 190–197.
13. Lodha, A.S., Pandya, A., Shukla, R.K. (2016) Nanotechnology: an applied and robust approach for forensic investigation. *Forensic Research & Criminology International Journal* 2(1): 00044.
14. Srividya, B. (2016) Nanotechnology in forensics and its application in forensic investigation. *Research Reviews Journal of Pharmacology and Nanotechnology* 4(2): 1–7.
15. El-Bialy, B.E.S., El-Borai, N.B., El-Latif, A.S.A., El-Gaber Mohamed, M.A. (2016) Biochemical and histopathological alterations as forensic markers of asphyxiated rats and the modifying effects of salbutamol and/or digoxin pretreatment. *Journal of Forensic Toxicology and Pharmacology* 5(1): 25–27.
16. Shanks, K.G., Winston, D., Heidingsfelder, J., Behonick, G. (2015) Case reports of synthetic cannabinoid XLR-11 associated fatalities. *Forensic Science International* 252.
17. Dlamini, B.C., Madala, N.E. (2017) Phytochemical composition and antioxidant and antimicrobial activities of *Solanum retroflexum* leaf extracts. Dissertation thesis retrieved from https://ujcontent.uj.ac.za/esploro/outputs/graduate/Phytochemical-composition-and-antioxidant-and-antimicrobial/9912302707691#file-0, accessed 30 October 2023
18. Aboubakr, M., Abdelazem, A.M., Abdellatif, A.M. (2014) Influence of *Aeromonas hydrophila* infection on the disposition kinetic of norfloxacin in goldfish (*Carassius auratus auratus*). *Journal of Forensic Toxicology and Pharmacology* 3: 1–5.
19. Maroof, K., Zafar, F., Ali, H., Naveed, S., Tanwir, S. (2016). Scope of nanotechnology in drug delivery. *Journal of Bioequivalence & Bioavailability* 8: 1–5.
20. Colton, R.J., Russell Jr, J.N. (2003) Making the world a safer place. *Science* 299(5611): 1324–1325.
21. Kosal, M.E. (2020) *Disruptive and Game Changing Technologies in Modern Warfare.* Springer International Publishing.
22. Mosher, P.V., McVicar, M.J., Randall, E.D., Sild, E.H. (1998) Gunshot residue-similar particles produced by fireworks. *Canadian Society of Forensic Science Journal* 31(3): 157–168.
23. Chalmers, J.M., Edwards, H.G., Hargreaves, M.D. (2012) Vibrational spectroscopy sampling techniques. *Infrared and Raman Spectroscopy in Forensic Science* 45–86.
24. López, T., Jardon, G., Gomez, E., Gracia, A., Hamdan, A., LuisCuevas, J., … Novaro, O. (2015) Ag/TiO$_2$-SiO$_2$ sol gel nanoparticles to use in hospital-acquired infections (HAI). *Journal of Materials Science and Engineering* 4: 196.
25. Abdellatif, A.A. (2015) Targeting of somatostatin receptors using quantum dots nanoparticles decorated with octreotide. *Journal of Nanomedicine & Nanotechnology* (S6): 1.
26. Catanzaro, M. (2021) Nanotechnology on the crime scene. *Nano Futures,* access from www.theguardian.com/what-is-nano/nanotechnology-on-the-crime-scene
27. Tefas, L.R., Tomuță, I., Achim, M., Vlase, L. (2015) Development and optimization of quercetin-loaded PLGA nanoparticles by experimental design. *Clujul Medical* 88(2): 214.
28. Sahu, G., Nayak, S.R., Sethi, S.S. (2016) Mutilation of body after crime. *Journal of Forensic Toxicology and Pharmacology* 5:1.
29. Worley, C.G., Wiltshire, S.S., Miller, T.C., Havrilla, G.J., Majidi, V. (2006) Detection of visible and latent fingerprints using micro-X-ray fluorescence elemental imaging. *Journal of Forensic Sciences* 51(1): 57–63.

30. Awoyera, P.O., Adesina, A., Gobinath, R. (2019) Role of recycling fine materials as filler for improving performance of concrete-a review. *Australian Journal of Civil Engineering* 17(2): 85–95.

31. Menzel, E.R. (2001) Recent advances in photoluminescence detection of fingerprints. *The Scientific World Journal* 1: 498–509.

32. Razak, S., Afsar, T., Al-Disi, D., Almajwal, A., Arshad, M., Alyousef, A.A., Chowdary, R.A. (2020) GCMS fingerprinting, in vitro pharmacological activities, and in vivo anti-inflammatory and hepatoprotective effect of selected edible herbs from Kashmir valley. *Journal of King Saud University –Science* 32(6): 2868–2879.

33. Spindler, X., Hofstetter, O., McDonagh, A.M., Roux, C., Lennard, C. (2011) Enhancement of latent fingermarks on non-porous surfaces using anti-l-amino acid antibodies conjugated to gold nanoparticles. *Chemical Communications* 47: 5602.

34. Hudson, M., Stuchinskaya, T., Ramma, S., Patel, J., Sievers, C., Goetz, S., Hines, S., Menzies, E., Russell, D.A. (2019) Drug screening using the sweat of a fingerprint: lateral flow detection of Δ9-tetrahydrocannabinol, cocaine, opiates and amphetamine. *Journal of Analytical Toxicology* 43(2): 88–95.

35. Hussain, C.M., Rawtani, D., Pandey, G., Tharmavaram, M. (2021) *Atomic Force Microscopy for Forensic Samples. Handbook of Analytical Techniques for Forensic Samples*, pp. 259–279. Elsevier.

36. Stylianou, A., Kontomaris, S.V., Grant, C., Alexandratou, E. (2019) Atomic force microscopy on biological materials related to pathological conditions. *Scanning*.

37. Murali, J. (2019) Nanotechnology on the crime scene. *Deccan Chronicle*, Apr 1, 2019. www.deccanchronicle.com/nation/current-affairs/010419/nanotechnology-on-the-crime-scene.html.

38. Lele, A. (2009) Role of nanotechnology in defence. *Strategic Analysis* 33(2): 229–241.

39. Nadar, S.S., Kelkar, R.K., Pise, P.V., Patil, N.P., Patil, S.P., Chaubal-Durve, N.S., … Patil, P.D. (2021) The untapped potential of magnetic nanoparticles for forensic investigations: A comprehensive review. *Talanta* 230: 122297.

40. Chakraborty, D., Rajan, G., Isaac, R. (2015) A splendid blend of nanotechnology and forensic science. *Journal of Nanotechnology in Engineering and Medicine* 6(1): 010801.

41. Giannoukos, S., Brkic, B., Taylor, S., Marshall, A., Verbeck, G.F. (2016) Chemical sniffing instrumentation for security applications. *Chemical Reviews* 116(14): 8146–8172.

42. Upadhyayula, V.K. (2012) Functionalized gold nanoparticle supported sensory mechanisms applied in detection of chemical and biological threat agents: a review. *Analytica Chimica Acta* 715: 1–18.

43. García-Briones, G., Olvera-Sosa, M., Palestino, G. (2019) Novel supported nanostructured sensors for chemical warfare agents (CWAs) detection. In: *Nanoscale Materials For Warfare Agent Detection: Nanoscience for Security 1*, pp. 225–251. Springer Netherlands.

44. Yue, G., Su, S., Li, N., Shuai, M., Lai, X., Astruc, D., Zhao, P. (2016) Gold nanoparticles as sensors in the colorimetric and fluorescence detection of chemical warfare agents. *Coordination Chemistry Reviews* 311: 75–84.

45. Martí, A., Costero, A.M., Gaviña, P., Parra, M. (2014) Triarylcarbinol functionalized gold nanoparticles for the colorimetric detection of nerve agent simulants. *Tetrahedron Letters* 55(19): 3093–3096.

46. Virel, A., Saa, L., Pavlov, V. (2009) Modulated growth of nanoparticles. Application for sensing nerve gases. *Analytical Chemistry* 81(1): 268–272.

47. Pena-Pereira, F., García-Figueroa, A., Lavilla, I., Bendicho, C. (2020) Nanomaterials for the detection of halides and halogen oxyanions by colorimetric and luminescent techniques: a critical overview. *TRAC Trends in Analytical Chemistry* 125: 115837.

48. Deng, X., Li, W., Wang, Y., Ding, G. (2020) Recognition and separation of enantiomers based on functionalized magnetic nanomaterials. *TRAC Trends in Analytical Chemistry* 124: 115804.

49. Ferreira, P.C., Ataide, V.N., Chagas, C.L.S., Angnes, L., Coltro, W.K.T., Paixão, T.R.L.C., de Araujo, W.R. (2019) Wearable electrochemical sensors for forensic and clinical applications. *TrAC Trends in Analytical Chemistry* 119: 115622.

50. Huang, X., Zhou, H., Huang, Y., Jiang, H., Yang, N., Shahzad, S. A., ... Yu, C. (2018) Silver nanoparticles decorated and tetraphenylethene probe doped silica nanoparticles: a colorimetric and fluorometric sensor for sensitive and selective detection and intracellular imaging of hydrogen peroxide. *Biosensors and Bioelectronics* 121: 236–242.

51. Sun, J., Zhang, X., Li, T., Xie, J., Shao, B., Xue, D., ... Liu, Y. (2019) Ultrasensitive on-site detection of biological active ricin in complex food matrices based on immunomagnetic enrichment and fluorescence switch-on nanoprobe. *Analytical Chemistry* 91(10): 6454–6461.

52. Pandya, A., Shukla, R. K. (2018) New perspective of nanotechnology: role in preventive forensic. *Egyptian Journal of Forensic Sciences* 8(1): 1–11.

53. Peveler, W.J., Jaber, S.B., Parkin, I.P. (2017) Nanoparticles in explosives detection—the state-of-the-art and future directions. *Forensic Science, Medicine and Pathology* 13(4): 490–494.

54. Rawtani, D., Tharmavaram, M., Pandey, G., Hussain, C.M. (2019) Functionalized nanomaterial for forensic sample analysis. *TrAC Trends in Analytical Chemistry* 120: 115661.

55. Ye, X., Wang, H., Yu, L., Zhou, J. (2019) Aggregation-induced emission (AIE)-Labeled cellulose nanocrystals for the detection of nitrophenolic explosives in aqueous solutions. *Nanomaterials* 9(5): 707.

56. Senesac, L., Thundat, T.G. (2008) Nanosensors for trace explosive detection. *Materials Today* 11(3): 28–36.

57. Nawaz, M.A.H., Meng, L., Zhou, H., Ren, J., Shahzad, S.A., Hayat, A., Yu, C. (2021) Tetraphenylethene probe based fluorescent silica nanoparticles for the selective detection of nitroaromatic explosives. *Analytical Methods* 13(6): 825–831.

58. Mauricio, F.G.M., Silva, J.Y.R., Talhavini, M., Júnior, S.A., Weber, I.T. (2019) Luminescent sensors for nitroaromatic compound detection: Investigation of mechanism and evaluation of suitability of using in screening test in forensics. *Microchemical Journal* 150: 104037.

59. Dasary, S.S., Singh, A.K., Senapati, D., Yu, H., Ray, P.C. (2009) Gold nanoparticle based label-free SERS probe for ultrasensitive and selective detection of trinitrotoluene. *Journal of the American Chemical Society* 131(38): 13806–13812.

60. Ganesan, M., Nagaraaj, P. (2020) Quantum dots as nanosensors for detection of toxics: a literature review. *Analytical Methods* 12(35): 4254–4275.

61. Sametband, M., Shweky, I., Banin, U., Mandler, D., Almog, J. (2007) Application of nanoparticles for the enhancement of latent fingerprints. *Chemical Communications* 11: 1142–1144.

62. Prasad, V., Lukose, S., Agarwal, P., Prasad, L. (2020) Role of nanomaterials for forensic investigation and latent fingerprinting—a review. *Journal of Forensic Sciences* 65(1): 26–36.

63. Choi, M.J., McDonagh, A.M., Maynard, P., Roux, C. (2008) Metal-containing nanoparticles and nano-structured particles in fingermark detection. *Forensic Science International* 179(2–3): 87–97.
64. Lodha, A.S., Pandya, A., Shukla, R.K. (2016) Nanotechnology: an applied and robust approach for forensic investigation. *Forensic Research & Criminology International Journal* 2(1): 00044.
65. Choi, M.J., McDonagh, A.M., Maynard, P., Roux, C. (2008) Metal-containing nanoparticles and nano-structured particles in fingermark detection. *Forensic Science International* 179(2–3): 87–97.
66. Choi, M.J., Smoother, T., Martin, A.A., McDonagh, A.M., Maynard, P.J., Lennard, C., Roux, C. (2007) Fluorescent TiO_2 powders prepared using a new perylene diimide dye: Applications in latent fingermark detection. *Forensic Science International* 173(2–3): 154–160.
67. Sancey, L., Lux, F., Kotb, S., Roux, S., Dufort, S., Bianchi, A., … Tillement, O. (2014) The use of theranostic gadolinium-based nanoprobes to improve radiotherapy efficacy. *The British Journal of Radiology* 87(1041): 20140134.
68. Brandão, M.D.S., Jesus, J.R., de Araújo, A.R., de Carvalho, J.G., Peixoto, M., Plácido, A., … Montagna, E. (2020) Acetylated cashew-gum-based silver nanoparticles for the development of latent fingerprints on porous surfaces. *Environmental Nanotechnology, Monitoring & Management* 14: 100383.
69. Goubko, M.V., Perepechina, I.O. (2015) Mathematical algorithm and the poll to reveal the criteria of the socially acceptable balance in judicial decisions taken using probabilistic data. *Journal of Forensic Research* 6(87).
70. Lad, A.N., Pandya, A., Agrawal, Y.K. (2016) Overview of nano-enabled screening of drug-facilitated crime: a promising tool in forensic investigation. *TrAC Trends in Analytical Chemistry* 80: 458–470.
71. Wu, Z., Zhou, H., Han, Q., Lin, H., Han, D., Li, X. (2020) A cost-effective fluorescence biosensor for cocaine based on a "mix-and-detect" strategy. *Analyst* 145(13): 4664–4670.

9 Understanding the Mechanism of Microbe-Mediated Nanosynthesis
A Molecular Approach

Smita Badur Karmankar, Swati Mehra,
Alka Sharma, and Ranjana Ahirwar Choudhary

9.1 INTRODUCTION

Over the last two decades, the field of nanotechnology has achieved some remarkable advances, transforming early promise into reality and hence strengthening the economy and improving quality of life. The direct implementations of nanotechnologies in the field of medical sciences have been a significant breakthrough for the scientific community.

Recent decades have witnessed the importance of nanomaterials (NMs) due to their increasing biomedical and industrial exploration. Particles possessing a size of 10–1,000 nm are regarded as nanoparticles [1]. Because of the high surface-to-volume ratio, NMs reflect unique and significant transformations regarding physical, chemical, and biological characteristics in comparison to bulk substances of the same chemical compounds. The notable features of nano-scale materials are their small size (1–100 nm), large specific surface area, high reactivity, and quantum effect [2–6]. Because of their distinctive structural parameters, NMs are highly explored in various areas such as materials science, biomedical engineering, and environmental engineering [7–15].

Due to simple penetration and identical sizes comparable to biomolecules, generally <100 nm is the most appropriate size for applications. Biologists have explored several research opportunities because of the smaller size of nanomaterials. Nanomaterials possess the potential to interact with complex biological systems in specific pathways because the dimensions of NMs are comparable to those of biomolecules. This rapidly growing area has provided the opportunity to design and synthesize multipurpose nanoparticles to recognize and cure diseases [16,17]. Recently, new approaches to drug delivery via nanocarriers are being explored for controlled and targeted delivery to specific sites and hence upgrading the efficacy of drugs and reducing the side effects of drugs during treatment [16,18,19]. Moreover,

DOI: 10.1201/9781003362258-9

nanocarriers interact with the biomolecules inside the cell and on the cell surface in such a way that the biochemical nature and characteristics of molecules are not changed [16,20,21]. Currently, nanomaterials having specific optical properties, such as surface plasmon resonance (SPR) and fluorescence, are gaining considerable attention in biomedical applications [22–24], particularly in generating optics-based analytical techniques used for bioimaging [25,26] and biosensing [27–29]. To acquire the aforementioned biomedical applications, the biocompatibility of a metal surface is a salient feature. In this context, metal nanoparticles are generated by exploiting biological systems; consequently metal ions are synthesized with high biocompatibility.

Numerous noticeable applications of nanoparticles in the field of biological and medicinal sciences, such as drug delivery, biomolecule detection, antimicrobial coatings, and diagnosis, are available [30]. Notably, nanoparticles are currently showing a crucial role in various mRNA deliveries, including both the Moderna and Pfizer-BioNTech COVID-19 vaccines [31–34].

The applications of nanomaterials in the field of environmental sciences are also considerable. Nanotechnologies are used in devices like solar cells which provide clean energy.

Nanoparticles like CNTs and fullerenes are generated intentionally using specific processes and hence are regarded as manufactured or engineered nanoparticles. In the context of environmental concerns, one-dimensional (1D) or two-dimensional (2D) systems can be used in chemistry, electronics, and engineering applications as thin films in a range of sizes (1–100 nm) in the field of catalysis or solar cells. These specialized thin films are used in various technological applications such as the development of an advanced system of chemical and biological sensors, environmental sensing systems, optical device, fiber-optic systems, and magneto-optics.

In addition, nanomaterials have a great impact on daily life and are now successively commercialized as commodity products [35]. These types of daily products have been used in the environment, agriculture, food, and medicine sectors [36]. With the continuous increase in quality of life, nanotechnology is being used in the production of daily use products [37], such as textiles, sports, cosmetics, and home appliances [38].

Due to the increased applications of NMs in various walks of human life, serious efforts toward the development of novel and innovative synthetic methods are highly recommended. Generally, two types of traditional synthesis protocols, physical and chemical strategies, are available. Common examples of physical methods are the ball-milling method, crushing method, and sputtering method [39–41]. However, the drawbacks associated with the physical method are the high energy consumption and long manufacture time. Due the mechanical obstructions, it is not easy to produce ultra-fine particles having stable features [42]. Typical examples of chemical synthesis are the solvothermal method, sonochemical method, sol–gel method, and microemulsion method [43–47]. However, chemical methods are considered as common synthetic routes for the production of NMs, but in chemical methods many toxic volatile chemicals and/or by-products are produced, and consequently human health and the environment are negatively affected [48,49]. Hence, the development of sustainable synthetic protocols for the generation of NMs is highly desirable.

Therefore, the development of reliable, nontoxic, clean, ecofriendly, and green experimental protocols for the synthesis of NPs is highly desirable [50–56]. In this regard, the exploration of biological resources such as microorganisms, enzymes, microbial enzymes, polysaccharides, vitamins, and biodegradable polymers for the synthesis of NPs is underway.

A remarkable alternative to traditional methods for the synthesis of NMs is the exploration of bio-mediated synthetic strategies using microbes [57]. Microbial synthesis is considered to be an environmentally benign sustainable strategy that uses biological creatures such as bacteria, actinomycetes, algae, fungi, viruses, and yeast for the production of NMs. The microbe-mediated strategy offers a non-hazardous, cost-effective, and authentic pathway for the synthesis of NMs. With the help of the aforementioned strategy, NMs are obtained in diverse sizes, shape, composition, and physicochemical characteristics. This sustainable method for the synthesis of NMs is regarded as a favorable application that enables production in an aqueous medium with minimum investment and low energy requirement, and can be easily maintained up to a large scale [58]. The extra advantage of these microbiological agents is their potential to act as templates for the production and arrangement of nano-range particles in a well-arranged pattern. However, over the past decade, a number of studies have reported on the biosynthesis of nanoparticles, and this chapter emphasizes a review of the current trends and progression achieved in NM production. This review chapter focuses on the exploitation of several microbes for the synthesis of metal nanoparticles, as well as knowledge regarding the plausible mechanisms suspected in the fabrication of metal nanoscale particles. In addition, the chapter discusses several applications and advantages of microbial approaches in nanoparticle synthesis. Finally, recent breakthroughs in large-scale production and concluding remarks the regarding future potential are also summarized.

9.2 MICROBE-MEDIATED SYNTHESIS OF METAL NANOPARTICLES

In the last decade, microbe-mediated synthesis of NMs has been regarded as a breakthrough in the field of green and sustainable nanotechnology [57]. Microbes act as substantial nanofactories for environmentally friendly and cost-effective synthesis of various metallic nanoparticles such as copper, palladium, silver, gold, and metal oxides, for example, zinc oxide, titanium oxide, etc. These nanoscale structures can be present in various forms and shapes such as nanorods, nanowires, nanotubes, nanoconjugates, etc. (Figure 9.1) [59]. The aforementioned various morphological forms have demonstrated attractive properties for use in biomedical applications as antimicrobial and anticancer agents.

9.2.1 Bacteria and Actinomycetes

Bacteria have unique potential for the reduction of metallic ions into nanoparticles. Easy handling and high growth rates makes bacteria one of the most suitable agents for the synthesis of nanoparticles. In contrast to other microbes, bacteria can be

Microbes

(Bacteria, Fungi, Algae, Yeast, Virus)

Intracellular or extracellular
extraction

Green-extract: →
- Proteins
- Polysaccharides
- Nitrate reductase
- Co-enzymes
- Bio-surfactant

+

Metal solution

Green synthesis

- Nanorods
- Nanowires
- Nanotubes
- Nanoconjugates

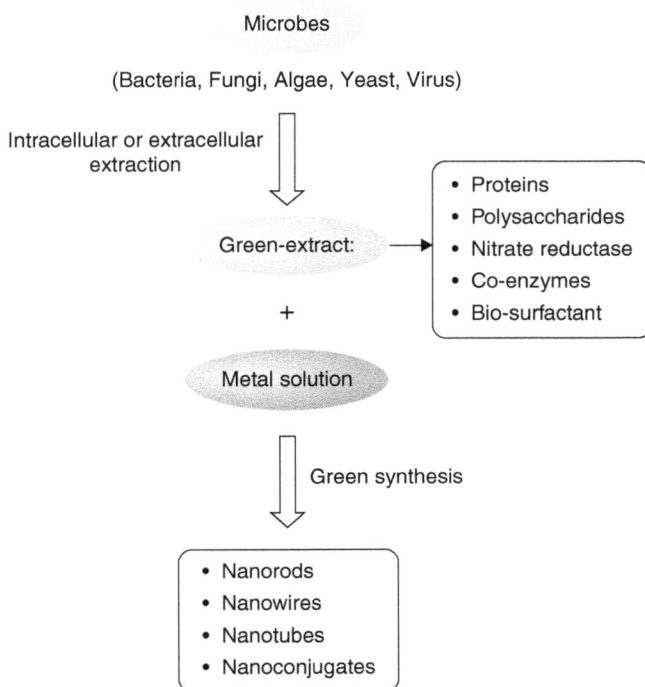

FIGURE 9.1 A pictorial representation of the microbe-assisted production of metal nanoparticles.

simply molded and genetically manipulated to biomineralize metal ions [60]. Bacteria are constantly exposed to toxic and hazardous environmental situations because of high concentrations of heavy metal ions in their surroundings. However bacteria have developed several natural defense mechanisms like efflux pumps, intracellular sequestration, change in metal ion concentration, and extracellular precipitation, to adjust to these adverse conditions [61]. These types of defense strategies could be potentially used by bacteria to produce nanoparticles. A list of some important bacterial strains exploited for the fabrication of NMs is provided in Table 9.1.

Generally, bacteria are considered to generate metal nanoparticles using intracellular or extracellular mechanisms. For the first time, Beveridge and Murray studied the deposition of gold nanoparticles (AuNPs) in an extracellular manner on *Bacillus subtilis* cell wall when a gold chloride solution was utilized for the suspension of an unfixed wall [62]. Next, a silver-resistant strain of *Pseudomonas stutzeri* AG259 deposited silver nanoparticles (AgNP) in intracellular fashion with the help of NADH-dependent reductase enzyme to produce NMs having a size range of a few nm to 200 nm [63]. Srivastava et al. discovered that *Pseudomonas aeruginosa* has the potential to produce different types of nanoparticles intracellularly, such as Li, Fe, Co, Ni, Ag, Pd, Pt, and Rh nanoparticles [64]. No external stabilizing agents and electron

TABLE 9.1
A Representative List of Bacteria and Actinomycetes Used for the Synthesis of Nanoparticles and Their Applications

Bacteria	Source of isolation	Metal nanoparticles	Size (nm)	Shape	References
Ochrobactrum rhizosphaerae	Marine water	Silver	10	Spherical	76
Bacillus sp. CS11	Soil samples from Cochin, India	Silver	42–92	Spherical	65 21
Deinococcus radiodurans	American Type Culture Collection, Manassas, USA	Silver	4–50	Spherical	66
Shewanella loihica PV-4	DSMZ, Germany	Palladium and platinum	2–7	Spherical	67
Ochrobactrum sp. MPV1	Roasted arsenopyrites, Tuscany, Italy	Tellurium	nd	Roughly spherical and rods	68
Bacillus subtilis	Hutti gold mine, India	Gold	20–25	Spherical	69
Bacillus brevis NCIM 2533	National Collection of Industrial Microorganism (NCL), Pune, India	Silver	41–68	Spherical	
Stenotrophomonas GSG2	Coral sample collected from Bay of Bengal	Silver and gold	Gold: 10–50; Silver: 40–60	Circular, triangular, hexagonal	74
Kocuria flava	Kanyakumari coast of India	Copper	5–30	Spherical	75
Alteromonas macleodii	Sediment sample from Kochi back water, India	Silver	70	Spherical	76
Pseudomonas aeruginosa JP-11	Marine water	Cadmium sulfide	20–40	Spherical	77
Bacillus cereus	Leaf of *Garcinia xanthochymus*	Silver	20–40	Spherical	78
Alcaligenes faecalis	Coral from Palk Bay located near Mandapam, Gulf of Mannar	Silver	30–50	Spherical	79

Actinomycetes					
Rhodococcus NCIM 2891	nd	Silver	10	Spherical	84
Streptomyces sp. LK3	Marine soil sample, Nicobar Island	Silver	5	Spherical	85
Streptacidiphilus durhamensis	Acidic forest soil	Silver	8–48	Spherical	86
Streptomyces rochei MHM13	Sediment samples along Suez Gulf, Red Sea, Egypt	Silver	22–85	Spherical	87
Streptomyces griseoruber	Soil Sample, Mercara region	Gold	5–50	Spherical	88
Streptomyces capillispiralis Ca-1	Medicinal plant Convolvulus arvensis	Copper	3.6–59	Spherical	89
Rhodococcus sp. NCIM 2891	National Chemical Laboratory, India	Silver	10–15	Spherical	90

donors were used in this work and this can be achieved without modifying the pH during the biomineralization step of various metal ions. Recently, several bacterial strains such as *Bacillus subtilis*, *Escherichia coli*, *Pseudomonas aeruginosa*, *Bacillus megaterium*, *Bacillus cereus*, *Klebsiella pneumoniae*, *Alteromonas*, *Ochrobactrum*, etc. have been frequently exploited for the synthesis of nanoparticles (Table 9.1).

Das et al. reported silver nanoparticle synthesis in an extracellular manner at a normal temperature in 24 h with the help of *Bacillus cereus* and the isolation of these bacteria was done from heavy-metal-contaminated soil [65]. The synthesized silver nanoparticles possessed surface plasmon resonance characteristics which may have broad applications. Kulkarni et al. reported radiation-resistant *Deinococcus radiodurans*-mediated reduction of silver chloride solution for the synthesis of AgNPs in an extracellular fashion [66]. Notably, AgNPs expressed broad-spectrum antibacterial activity against both Gram-positive and Gram-negative bacteria. In addition, AgNPs demonstrated very good anticancerous activity also against human breast cancer cell lines. On the basis of cytotoxicity assay and cell viability, it has been concluded that AgNPs have the potential to check the growth of cancer cell lines.

Recently, scientists have shown interest in the design and synthesis of various kinds of NMs such as tellurium, palladium, platinum, etc. In this context, Ahmed et al. investigated the preparation of ultra-small palladium and platinum nanoparticles using *Shewanella loihica* PV-4 having a size range of from 2 nm to 7 nm [67]. In this study, electrochemical-activated biofilms of *S. loihica* were exploited to synthesize ultra-small nanoparticles. Moreover, the generated palladium and platinum NMs expressed very good catalytic efficacy for the degradation of methyl orange dye. Zonaro et al. reported the production of tellurium NMs with the help of *Ochrobactrum* sp., and also described that this strain can be used as a remarkable nanofactory for the transformation of toxic tellurite oxyanions into practical nanoparticles [68]. Recently, Srinath et al. reported the preparation of AuNPs exploiting *Bacillus subtilis* [69]. The advantages of microorganisms extracted from gold mines include good resistance to toxicity associated with gold ions and they can produce AuNPs successfully. In addition, AuNPs are explored in the decomposition of methylene blue as a biocatalyst. Saravanan et al. reported the exploration of *Bacillus brevis* for the preparation of spherical silver nanoparticles within a size range of 41–62 nm [70]. Additionally, AgNPs expressed potential antibacterial activity against multidrug-resistant strains of *Staphylococcus aureus* and *Salmonella typhi*.

Various parameters are responsible for the reduction of metallic ions into nanoparticles. In this regard, the first remarkable parameter is the presence of functional organic molecules on the cell wall which causes biomineralization, and another important parameter is appropriate reaction conditions such as the constitution of the medium, temperature, pH, and concentration of metallic salt [63]. The aforementioned reaction conditions can extensively affect the size, composition, and morphology of nanoparticles [71]. Hence, the optimization of these reaction parameters is highly desirable during biosynthesis to enhance the overall quality of particles. In this regard, Ramanathan et al. reported the synthesis of silver nanoparticles (AgNPs) with the help of *Morganella psychrotolerans* and they also carried out optimization of the growth kinetics parameters to examine their impact on AgNP morphology [72]. It was observed that spherical AgNPs were synthesized with an average size range of 2–5 nm at an optimum temperature of 20°C, whereas a mixture of spherical, hexagonal, and triangular nanoparticles and nanoplates were generated at 25°C. On decreasing the temperature from 20°C to 15°C, again a mixture of spherical particles and nanoplates was produced. Next, the retardation in growth and physiological activity of bacteria was noticed when the optimum growth temperature was decreased to 4°C, consequently there was an extensive increase in the numerous nanoplates with only a small number of spherical nanoparticles. Additionally, at 4°C, the spherical nanoparticles were produced in comparatively larger size ranges from 70 nm to 100 nm. In another report, Yumei et al. investigated the synthesis of AgNPs with the help of *Arthrobacter* sp., and also pointed out that nanoparticle synthesis can be regulated by altering temperature, pH, and metal ion concentration [73]. Face-centered-cubic AgNPs with sizes between 9–72 nm can be obtained from low concentrations of silver nitrate (1 mM), pH 7–8, and 70°C temperature. As the concentration of silver nitrate was increased up to 3 mM, AgNPs were obtained in aggregated form at 70°C. Notably, the synthesis of AgNPs was not obtained below 5 or above 8 pH. As the incubation temperature was increased from 70°C to 90°C, the synthesis time was reduced from 10 min to 2 min. On the basis of the above discussion it can be concluded that moderate pH as well as metal ion concentration have direct impacts on the synthesis of nanoparticles.

In addition to terrestrial bacteria, recently, marine microbial cultures have also been extensively exploited as nanofactories to synthesize nanoparticles. Malhotra et al. examined the efficacy of a novel marine bacterium, *Stenotrophomonas*, for the production of AgNPs and AuNPs [74]. It was observed that secretory proteins of low molecular weight present in the supernatant were responsible for the biosynthesis of AuNPs and AgNPs. In a similar way, Kaur et al. reported a new marine strain, *Kocuria flava*, which has the potential to synthesize copper nanoparticles with particle sizes within the range of 5–30 nm [75]. A number of literature reports have suggested that effective bio-reductant and capping agent-mediated synthesis of nanoparticles was done with the help of bacterial extracellular polymeric substances (EPSs) [76–78]. Mehta et al. showed the exploration of EPS, secreted *Alteromonas macleodii*, for the synthesis of silver nanoparticles having a narrow size distribution [77]. Table 9.1 presents an overview of the outcomes of bacteria-mediated biosynthesis of nanoparticles [65–82].

Actinomycetes have been generally used for the synthesis of extracellular enzymes and secondary metabolites [83]. They have also been adopted for the biosynthesis of nanoparticles as they have unsurpassed capacity for the production of various bioactive compounds and have a high protein content. Actinomycetes synthesize nanoparticles via both intracellular and extracellular pathways, but extracellular reduction is the most common pathway and has more commercial applications in different fields. In 2012, Otari et al. explained the green biosynthesis of silver nanoparticles using actinobacteria *Rhodococcus* NCIM 2891 [84]. The TEM graph analysis of AgNPs revealed a spherical shape with an average diameter of 10 nm. Intracellular biomineralization of silver ions was thought to be the result of enzymes present on the cell wall, resulting in the production of silver nuclei. Karthik et al. adopted the marine bacterium, *Streptomyces* sp. LK-3, for the reduction of silver ion into AgNPs [85]. Their study concluded that nanoparticles were synthesized extracellularly and NADH-dependent nitrate reductase was mainly responsible for the reduction of silver ions into stable AgNPs via an electron transfer reaction. The AgNPs exhibited strong acaricidal or antiparasitic activity against *Rhipicephalus microplus* and *Haemaphysalis bispinosa*. Recently, Buszewski et al. employed an acidophilic actinobacteria, *Streptacidiphilus durhamensis*, for the synthesis of silver nanoparticles [86]. Their work illustrated the formation of stable spherical AgNPs within a size range of 8–48 nm which showed antibacterial activity against *Pseudomonas aeruginosa*, *Staphylococcus aureus*, and *Proteus mirabilis*. Generally, biosynthesized nanoparticles exhibit higher antimicrobial activity in comparison to traditionally synthesized nanoparticles due to the action of various bioactive molecules involved in capping and stabilization of the nanoparticles. Later, Abd-Elnaby et al. [86] screened actinomycetes isolates from the Suez Gulf, Red Sea, and found that only two strains were capable of synthesizing AgNPs [87]. Moreover, AgNPs exhibited strong antibacterial activity against various pathogenic bacteria such as *Pseudomonas aeruginosa*, *Escherichia coli*, *Bacillus subtilis*, *Staphylococcus aureus*, *Salmonella typhimurium*, *Vibrio damsela*, *Vibrio fluvialis*, and *Bacillus cereus*. It can be concluded that AgNPs remain the most widely studied nanoparticles by actinomycetes. However, there have been recent reports which have described the synthesis of copper and gold nanoparticles using *Streptomyces griseoruber* and *Streptomyces capillispiralis* Ca-1, respectively [88,89]. Among actinomycetes, species of *Streptomyces* are most widely used in pharmaceutical and enzymatic applications because, out of more than 10,000 known antibiotics, 55% are produced by them. Table 9.1 provides a list of various recent reports on nanoparticle synthesis by actinomycetes and their applications [84–92].

9.2.2   Fungi and Yeast

Fungal biosynthesis of nanoparticles is another simple and straightforward approach which has been explored extensively for the fabrication of nanoparticles. In comparison to bacteria, fungi have higher productivity in terms of nanoparticle generation and greater tolerance to metals, especially in context of high cell wall-binding capacity of metal ions with biomass [57]. The downstream processing and biomass treatments are relatively easy in fungi as compared to bacteria and viruses. Moreover, fungi possess higher bioaccumulation ability toward metal ions, resulting in the efficient and cost-effective production of nanoparticles. However, the process parameters have a significant effect on the biosynthesis of nanoparticles. An in-depth investigation of different process parameters was carried out by Bhargava et al. to study the effects of pH, salt concentration, and reaction time on the particle size and yield of fungi *Cladosporium oxysporum* to convert gold ions into nanoparticles [93]. The maximum yield of AuNPs was obtained with a biomass to water ratio of 1:5 at 1 mM salt concentration and pH 7. Moreover, the synthesized AuNPs exhibited excellent catalytic activity in the degradation of textile dye, rhodamine B, within 7 min. Mishra et al. also described the extracellular formation of gold nanoparticles by culture filtrate of *Hypocrea lixii* and *Trichoderma viride*, and studied the effects of reaction temperature and incubation time on nanoparticle biosynthesis [94]. *T. viride* had AuNP biosynthesis within 10 min at 30°C, which further served as a biocatalyst and strong antimicrobial agent. Metuku et al. collected a white rot fungus, *Schizophyllum radiatum*, from Eturnagaram forest of Warangal, India, and found it capable of producing well-dispersed stable silver nanoparticles [95]. Their research work investigated the potential of white rot fungus in the extracellular biomineralization of silver ions to nanoparticles of size 10–40 nm. These small-sized AgNPs demonstrated high antibacterial activity against various pathogenic Gram-negative and Gram-positive bacterial strains. Most of the studies reported till date have described the involvement of extracellular components in the fabrication of nanomaterials. The main advantage of extracellular-mediated nanoscale material synthesis is that it avoids impurities such as intracellular proteins, and treatment with detergents and ultrasound is not required.

In addition, understanding the mechanistic aspects of nanoparticle synthesis has also become indispensable for developing reliable applications. To overcome this knowledge gap, Rajput et al. explored various fungal strains of *Fusarium oxysporum* for silver nanoparticle synthesis and studied the effects of isolate selection, temperature, and pH on nanoparticle morphology [96]. Their study summarized that understanding the interactions between organic and interfacial layers will be helpful in developing novel uses, mainly in the area of biosensors.

To further explore the bioinspired formation of nanoparticles, Kitching et al. extracted the cell surface proteins of *Rhizopus oryzae* for the in vitro production of gold nanoparticles for biomedical and biocatalytic applications [97]. In 2017, Suryavanshi et al. explored the synthesis of aluminum oxide nanoparticles using *Colletotrichum* sp., and nanoparticles were functionalized by essential oils extracted from *Eucalyptus globulus* and *Citrus medica* [98]. The results concluded that nanofunctionalized oil can be used as antimicrobial agents against foodborne pathogens for the prevention of food spoilage. Recently, two filamentous fungi, *Penicillium citreonigrum* and *Scopulariopsis brumptii*, and an edible mushroom, *Pleurotus ostreatus*, have also been adopted for the synthesis of nanoparticles for anticancer and antimicrobial applications, respectively [99,100]. Table 9.2 provides a list of various fungi that have been used for the biosynthesis of different metallic nanostructures for various applications [93–111].

In addition to fungi, some researchers have investigated the use of yeasts for the biogenic synthesis of the nanoparticles. Yeasts possess the inherent capability to absorb and accumulate high concentrations of toxic metal ions from their surroundings [112]. Yeast cells adapt themselves under metal toxicity conditions using various detoxification mechanisms, namely bio-precipitation, chelation, and intracellular sequestration. This property of yeast cells has been exploited by various researchers. For example, in one study a marine strain of ascomycetous yeast *Yarrowia lipolytica* was employed for the biomimetic synthesis of silver nanoparticles in a cell-associated manner [113]. The study concluded that possibly brown pigment (melanin) obtained from the yeast cells was responsible for biomineralization of metal ions. The pigment-derived silver nanoparticles displayed antibiofilm activity against a *Salmonella paratyphi* pathogen. In another research work, Waghmare et al. reported the ecofriendly extracellular biosynthesis of AgNPs using *Candida utilis* NCIM 3469 [114]. These nanoparticles were circular in shape with a size of between 20–80 nm, and showed antibacterial activity against pathogenic strains, i.e. *Staphylococcus aureus, Pseudomonas aeruginosa,* and *Escherichia coli.* In a recent study, Elahian et al. utilized a genetically modified yeast, *Pichia pastoris,* for the biosynthesis of silver nanoparticles [115]. An engineered *Pichia pastoris* strain overexpressed a metal-resistant gene, cytochrome b5 reductase enzyme obtained from *Mucor racemosus,* for the reduction of metal ions into nanoparticles. The cytochrome b5 reductase enzyme leads to the synthesis of stable and well-dispersed metal nanoparticles within the size range of 70–180 nm. In 2016, Eugenio et al. isolated a yeast strain, *Candida lusitaniae,* from the gut of a termite and demonstrated the production of silver nanoparticles with diameters in the range of 2–10 nm [116]. The silver nanoparticles showed antiproliferative activity against *S. aureus* and *Klebsiella pneumoniae,* and presented a promising alternative to commonly used antibiotics. Sriramulu and Sumathi (2018) employed *Saccharomyces cerevisiae* aqueous extract for the synthesis of hexagonal palladium nanoparticles (PdNPs) of size 32 nm [117]. PdNPs showed photocatalytic degradation of textile azo dye (direct blue 71) to 98% within 60 min under UV light. All these literature reports suggested that the differences in nanoparticle size, shape, and properties are because of the different mechanisms adopted by yeast cells to synthesize and stabilize the nanoparticles. Table 9.2 provides a list of various yeasts used for the synthesis of different metal nanoparticles [113–130].

9.2.3 ALGAE

Similarly to yeast, there are diverse literature reports on algae being used as a "nanofactory" for the biosynthesis of metal nanoparticles (Table 9.3). Ferreira et al. employed the dried unicellular microalgae, *Chlorella vulgaris,* for the biosynthesis of silver nanoparticles within the size range of 5.7–9.8 nm [131]. The spherical-shaped nanoparticles were observed to be a promising green alternative for biomedical application as antimicrobial agents. In another study, Arsiya et al. evaluated the synthesis of palladium nanoparticles using *Chlorella vulgaris* aqueous extract within 10 min [132]. TEM results revealed that the nanoparticles were circular and mono-dispersed in nature, having a size of 5–20 nm. This study for the first time reported the synthesis of palladium nanoparticles in a comparatively shorter time duration using *C. vulgaris.* The biosynthesis of palladium nanoparticles has also been reported using marine algae, *Sargassum bovinum,* which was isolated from the Persian Gulf area [133]. Dhas et al. explored the synthesis of silver chloride nanoparticles using the aqueous extract of marine algae, *Sargassum plagiophyllum* [134]. Recently, an economical green method has been reported for the synthesis of silver nanoparticles using a marine green algae, *Caulerpa racemosa* [135]. The synthesized nanoparticles exhibited remarkable catalytic activity toward the degradation of methylene blue. Ramakrishna et al. studied the synthesis of gold nanoparticles using aqueous extracts of brown algae, *Sargassum tenerrimum* and *Turbinaria conoides* [136]. The AuNPs displayed excellent biocatalytic activity in the degradation of aromatic nitro compounds and organic dyes. The metal nanoparticles of zinc oxide (ZnO) have also generated curiosity among researchers due to their

TABLE 9.2
A Representative List of Fungi and Yeast Used to Synthesize Metal Nanoparticles and Their Applications

Fungi/Yeast	Source of isolation	Metal nanoparticles	Size (nm)	Shape	Ref.
Fusarium oxysporum 405	Obtained from American Research Service, Washington, USA	Silver	10–50	Spherical	96
Rhizopus oryzae	National Collection of Industrial Microorganism (NCIM), Pune, India	Gold	16–43	Spherical and flower-like	97
Colletotrichum sp.	Amravati University, Amravati, India	Aluminum oxide	30–50	Spherical	98
Pleurotus ostreatus	Biotechnology Center, Cairo University, Egypt	Gold	10–30	Spherical and prism shaped	99
Penicillium citreonigrum	East of Lake Burullus, Egypt	Silver	6–26	Spherical	100
Penicillium diversum	Microbial Type Culture Collection, Chandigarh, India	Silver	10–50	Roughly spherical	101
Aspergillus foetidus	Kalyani Waste Water Centre, West Bengal India	Silver	20–40	Roughly spherical	102
Fusarium oxysporum sp. cubense JT1	Isolated from wilt-infected banana plants	Gold	22	nd	103
Trichoderma harzianum	Procured from College of Life Sciences, Gwalior, India	Cadmium sulfide	3–8	Spherical	104
Botrytis cinerea	Isolated from rotten grapes collected from Region IV, Chile	Gold	1–100	Triangular, spherical, hexagonal, pyramidal, decahedral	105
Nigrospora oryzae	nd	Gold	6–18	Cubic and spherical	106
Aspergillus terreus	Microbial Type Culture Collection, Chandigarh, India	Zinc oxide	28–63	Spherical	107
Curvularia lunata	Leaves of *Catharanthus roeus*	Silver	10–50	Spherical	108

(Continued)

TABLE 9.2 (Continued)
A Representative List of Fungi and Yeast Used to Synthesize Metal Nanoparticles and Their Applications

Fungi/Yeast	Source of isolation	Metal nanoparticles	Size (nm)	Shape	Ref.
Metarhizium anisopliae	T-Stanes & Company Limited, Tamil Nadu, India	Silver	28–38	Rod-shaped	109
Trichoderma harzianum	nd	Silver	20–30	Spherical	110
Fusarium oxysporum	National Institute of Genetic Engineering and Biotechnology (NIGEB), Tehran, Iran	Silver	34–44	Spherical	111
Yarrowia lipolytica NCYC789	National Collection of Yeast Cultures, Norwich, U.K.	Silver	15	nd	112
Candida utilis NCIM 3469	National Collection of Industrial Microorganism (NCIM), Pune, Maharashtra, India	Silver	20–80	Spherical	113
Pichia pastoris	Recombinant strain overexpressing *Mucor racemosus* cytochrome b5 reductase	Silver	70–180	Spherical	114
Candida lusitaniae	Isolated from gut of *Coruitermes cumulans* termite	Silver and silver chloride	2–10	Cubical, cuboctahedral, Icosahedral, and spherical	115
Saccharomyces cerevisiae	Purchased from Pagariya Food Products (P) Ltd, Tamil Nadu, India	Palladium	32	Hexagonal	116
Rhodosporidium diobovatum	Isolated from Indian Ocean	Lead sulfide	2–5	Spherical	117
Cryptococcus laurentii	nd	Silver	35	Roughly spherical	118
Saccharomyces cerevisiae	Purchased from local market	Silver	2–20	Spherical	119
Magnusiomyces ingens LHF1	Sea mud of Harbor Industrial Zone, Dalian, China	Gold	15	Spherical	120
Cryptococcus laurentii	Isolated from apple peel	Silver	15–35	Spherical	121

Organism	Source	Material	Size	Shape	Reference
Saccharomyces cerevisiae	AB Mauri (P) Ltd, Bangaluru, Karnataka, India	Gold nanoplates	–	Hexagonal and triangular nanoplates	122
Candida albicans ATCC 10231	American Type Culture Collection, Manassas, USA	Silver	10–20	Spherical	123
Pichia kudriavzevii	Isolated from Sourdoughs and Tanzanian Togwa	Zinc oxide	10–61	Hexagonal wurtzite structure	124
Rhodotorula mucilaginosa	Copper waste pond at the Sossego Mine, Brazil	Silver	11	Spherical	125
Phaffia rhodozyma	American Type Culture Collection (ATCC), USA	Silver and gold	Silver: 5–9, gold: 4–7	Spherical and quasispherical	126
Magnusiomyces ingens LHF1	Sea mud of Harbor Industrial Zone, Dalian, China	Gold	20.3–28.3	Spherical and pseudospherical	127
Rhodotorula glutinis	Soil sample of Pici Campus, The Federal University of Cear'a, Brazil	Silver	16	Spherical	128
Candida glabrata	Oropharyngeal mucosa of HIV patients	Silver	2–15	Spherical	129

unique physicochemical characteristics and wide applications in opto-electronics, sunscreens, biomedicines, food additives, etc. Rajeshkumar adopted two marine brown seaweeds, *Padina tetrastromatica* and *Turbinaria conoides*, in an algal formulation for the biosynthesis of ZnO nanoparticles and evaluated their antimicrobial potential against fish pathogens [137]. In another recent work, Sanaeimehr et al. synthesized ZnO nanoparticles using *Sargassum muticum* extract and established their antiangiogenic and antiapoptotic potential against human liver cancer cell lines [138]. All these literature reports indicate that researchers are now exploring marine organisms for the biogenic synthesis of nanoparticles because marine algae contain various biologically active compounds and secondary metabolites that allow them to act as "nanofactories" [139]. These marine algae have many applications in biomedicine as antioxidants, anticancer, antidiabetic, cardioprotective, hepatoprotective, and antiviral agents. Table 9.3 highlights the results of recent reports on the algae-based biosynthesis of metal nanoparticles for different biological applications [131–154].

9.2.4 Viruses

An interesting property of viruses is their thick outer surface coating of capsid proteins which provides a highly suitable platform for interaction with metallic ions [155]. These protein cages can build monodispersed units that are highly robust and moldable through genetic engineering. Viruses can be modified to serve as templates for material deposition or engineered to create three-dimensional vessels for targeted drug delivery [156]. Viruses can be employed for the synthesis of nanoconjugates and nanocomposites with metal nanoparticles which are important bioengineering materials in drug delivery and cancer therapy (Table 9.4). Mao et al. investigated the use of M13 bacteriophages for the nucleation and orientation process of semiconductor nanocrystals [157]. This group showed a genetically controllable biogenic synthesis route to semiconductor nanocrystals of zinc sulfide and cadmium sulfide. The plant viruses have been proved to be safe for nanotechnology applications due to their structural and biochemical stability, ease of cultivation, non-toxicity, and non-pathogenicity in animals and humans. In one study, low concentrations of tobacco mosaic virus (TMV) and bovine papilloma virus (BPV) were used as additives along with extracts of various plants, e.g. *Nicotiana benthamiana*, *Avena sativa*, and *Musa pradisiaca*, etc. [158]. The TMV and BPV not only helped in the reduction of size, but also significantly enhanced the numbers of nanoparticles in comparison to the non-virus control. Cao et al. employed red clover necrotic mosaic virus (RCNMV) for the synthesis of nanoparticles for the controlled delivery of doxorubicin drug for chemotherapy [159]. The unique morphology of RCNMV and structural changes in response to divalent cations removal help in doxorubicin infusion to the capsid through the surface pore formation mechanism. Le et al. investigated the potential of potato virus X nanoparticles for the delivery of doxorubicin drug for cancer treatment [160]. Potato virus X has the capability to synthesize elongated filamentous nanoparticles which exhibit enhanced tumor homing and penetration power in comparison to spherical ones. However, the synthesis of nanoparticles by viruses still faces various drawbacks such as involvement of the host organism for protein expression,

TABLE 9.3
A Representative List of Algae Used for the Synthesis of Nanoparticles and Their Applications

Algae	Source of isolation	Metal nanoparticles	Size (nm)	Shape	References
Chlorella vulgaris	Culture Collection of Algae, University of Texas, Austin	Silver	0.3–15	Spherical	131
Chlorella vulgaris	Faculty of Natural Resources and Environment, University of Birjand, Iran	Palladium	0.5–20	Spherical	132
Sargassum bovinum	Persian Gulf area, South Western, Iran	Palladium	0.5–10	Octahedral	133
Sargassum plagiophyllum	Rameshwaram Coast, Tamil Nadu, India	Silver	18–42	Spherical	134
Caulerpa racemosa	Mandapam Coastal Area, Tamil Nadu, India	Silver	25	Distorted spherical	135
Sargassum tenerrimum	Mandapam Coast, Tamil Nadu, India	Gold	27–35	Spherical	136
Padina tetrastromatica	Tuticorin Coast, Tamil Nadu, India	Zinc oxide	90–120	Spherical	137
Sargassum muticum	Northwest Pacific Region, Iran	Zinc oxide	30–57	Hexagonal	138
Porphyra vietnamensis	nd	Silver	13	Spherical	140
Stoechospermum marginatum	Tuticorin Coast, Tamil Nadu, India	Gold	19–94	Spherical, hexagonal, and triangular	141
Tetraselmis kochinensis	nd	Gold	05–35	Spherical and triangular	142
Chaetomorpha linum	Kanyakumari Coast, India	Silver	03–44	Clusters	143
Spirogyra varians	Sweet water areas, Kerman, Iran	Silver	35	Quasi-spherical	144
Scenedesmus sp.	CSIR – Institute of Minerals and Materials Technology, Bhubaneswar, India	Silver	15–20	Spherical crystalline	145
Chlorella vulgaris	Algal Culture Collection, Chennai, India	Gold	02–10	Spherical self-assembled cores	146
Ecklonia cava	Busan, South Korea	Gold	30	Spherical and triangular	147
Caulerpa racemosa	Gulf of Mannar, Southeast Coast, India	Silver	05–25	Spherical and triangular	148

(Continued)

TABLE 9.3 (Continued)
A Representative List of Algae Used for the Synthesis of Nanoparticles and Their Applications

Algae	Source of isolation	Metal nanoparticles	Size (nm)	Shape	References
Ulva lactuca	Coastal areas of Rameshwaram, Tamiladu, India	Silver	20–35	Cubical	149
Pithophora oedogonia	Freshwater pond in Hoogly, West Bengal, India	Silver	25–44	Cubical and hexagonal	150
Cystoseira baccata	Northwest coast of Spain	Gold	8.4	Spherical	152
Galaxaura elongata	Northwest coast of Red Sea shore	Gold	3.8– 77.1	Spherical, rods, hexagonal and triangular	151
Laminaria japonica	Local seaweed industry in Korea	Silver	31	Spherical to oval	153
Gelidium amansii	Coastal Belt of South Korea	Silver	27–54	Spherical	154

underdeveloped processes for synthesis, and limited research on large-scale application. Table 9.4 summarizes the results of recent reports on virus-based synthesis of metal nanoparticles and their applications [155–164].

9.3 LARGE-SCALE PRODUCTION OF BIOMEDIATED NANOPARTICLES

Microbial fermentation represents the state-of-the-art approach for large-scale production of nanoscale structures of different metals. In recent years, researchers have explored large-scale synthesis of nanoparticles using biogenic routes with a narrow size distribution [165,166]. In 2010, Moon and his group for the first time reported large-scale production of magnetic and metal-substituted magnetic nanoparticles using *Thermoanaerobacter* sp. TOR-39 [167]. This report concluded that magnetic nanoparticle production can be obtained in huge quantities at low cost, similar to traditional chemical synthesis. At the end, about 1 kg (wet weight) of Zn-substituted magnetites was obtained from 30 L fermentations. The magnetic nanoparticles have become the focus of recent research as they are a promising candidate for magnetic resonance imaging, bioremediation, data storage, and catalysis biosensor development, and they can be manipulated very easily under magnetic field influence. In another report, Moon et al. employed the same thermophilic strain, *Thermoanaerobacter* sp. TOR-39, for the extracellular synthesis of cadmium sulfide (CdS) nanoparticles [168]. The size of CdS crystallites was less than 10 nm and the process was easily scalable up to 24 L. Various factors such as biomass concentration, dosing amount, type of precursors used, and the basal medium composition

TABLE 9.4
A Summary of the Results of Recent Reports on Virus-Based Synthesis of Metal Nanoparticles and Their Applications

Virus	Nanoparticle type	Size (nm)	Shape	References
Tobacco mosaic virus (TMV)	Gold	5	Spherical	155
Cucumber mosaic virus	Nanoassemblies	29	Icosahedral	156
Red clover necrotic mosaic virus	Nanocarriers	36	Icosahedral	159
Potato virus X	Nanocarriers	13	Helical	160
Tobacco mosaic virus (TMV)	Palladium	2.9–3.7	Multiwalled carbon nanotubes	161
M13 virus	Titanium dioxide	20–40	Mesoporous nanowires	162
Potato virus X	Nanoconjugates	12	Filamentous rod shaped	163
Hepatitis E virus	Nanoconjugates	27–34	Icosahedral	164

were found to be crucial for producing tailor-made nanoparticles. Later, Moon et al. elucidated the synthesis of semiconducting zinc sulfide (ZnS) nanoparticles in a lab-scale reactor with 24 L capacity using an anaerobic thermophilic metal-reducing bacterium, *Thermoanaerobacter* sp. X513 [165].

The production of ZnS nanoparticles was scalable, reproducible, and controllable (within the 2–10 nm range) from 10 mL to 24 L, with yields of 5 g per L per month. More recently, Moon et al. demonstrated the scale-up of nanoparticle synthesis from the lab scale to pilot-plant level using the same bacterium [169]. This work investigated the scalability of bacteria-mediated ZnS nanoparticle production in 100 L and 900 L scale bioreactors. Repeated 100 L batches using fresh or recycled media produced ZnS nanoparticles with high reproducibility in a crystalline size of 2 nm with yields of approximately 0.5 g L^{-1} which were close to the small-scale batches. The cultivation at 900 L scale yielded around 320 g ZnS nanoparticle powder, and this amount was sufficient for the synthesis of a ZnS thin film with thickness of 120 nm over 0.5 m width and 13 km length. In another study, Ramos-Ruiz et al. discussed the potential of up-flow anaerobic sludge bed reactors for continuous conversion of toxic tellurite oxyanions (Te^{IV}) to non-toxic recoverable tellurium (Te^0) nanoparticles using methanogenic microbial consortium. This group also evaluated the effect of redox-mediating flavonoid compound, riboflavin (RF), with the aim of increasing the reduction of tellurite oxyanions. The presence of a riboflavin mediator enhanced the conversion rate of tellurite by approximately 11-fold. This work showed that the methanogenic anaerobic granular sludge can be adopted as a bioreactor technology for the continuous production of tellurium nanoparticles in direct recoverable mode. Moreover, the sludge was able to sustain the reduction of high loads of toxic tellurite oxyanions.

With respect to large-scale production, the fact that only few reports are available on bioreactor cultivation strategies for biosynthesis in last six years is not a good finding, as these strategies are essential for obtaining higher productivity of nanoparticle synthesis. These are important for large-scale production and thus need to be further investigated.

9.4 ADVANTAGES AND LIMITATIONS OF BIOLOGICAL METHODS IN NANOPARTICLE SYNTHESIS

There have been tremendous developments in the field of microorganism-produced nanoparticles and their applications over the last decade. The biosynthesis of nanoparticles has numerous advantages such as benign and eco-friendly production, cost-effectiveness, and the biocompatibility of synthesized nanoparticles [57]. As opposed to physicochemical processes, biosynthesized nanoparticles are free from toxic chemical contaminants which is essentially a desirable trait for biomedical applications [170]. Another benefit of the biogenic route of synthesis is that it does not require an additional step of capping or attachment of bioactive compounds to the nanoparticle surface to generate stable and pharmacological active particles, which is otherwise essential in physicochemical synthesis [73,76,85]. Furthermore, the time required for the biosynthesis of nanoparticles is much less than for the physicochemical methods. For example, Arsiya et al. demonstrated one-step biosynthesis of palladium nanoparticles using *Chlorella vulgaris* [132]. The reduction of palladium ions into nanoparticles was achieved within 10 minutes at room temperature. The FTIR analysis of *Chlorella vulgaris* extract revealed that polyol and amide groups present in extract act as reducing and stabilizing agents. Several other investigators have discovered rapid biosynthetic procedures with high nanoparticle yield using different algal extracts. For example, silver nanoparticles were synthesized by algal extracts within 2 min [152], 15 min [150], and 1 h [149]. Gold nanoparticles were also formed within 5 min [152] and 10 min [141], highlighting the importance of nanoparticle synthesis using biogenic agents. In spite of various advantages offered by the biological route for the synthesis, the polydispersity and size of the nanoparticles remain big and challenging issues. Further, much work is needed to improve the efficiency of synthesis, the control of particle size, and morphology. Thus, several recent reports have developed a stable system for nanoparticle biosynthesis with monodispersity in size and shape. The size and shape of metal nanoparticles could be controlled by either optimizing the process parameters or modifying these parameters. For example, Hamedi et al. demonstrated the synthesis of highly monodispersed silver nanoparticles using *F. oxysporum* by altering the process conditions such as incubation time, temperature, metal salt concentration, and C:N ratio [111]. The increase in C:N ratio resulted in the synthesis of small-sized AgNPs with high monodispersity and productivity. In 2018, Domany et al. synthesized stable gold nanoparticles with moderate dispersity using *Pleurotus ostreatus* extracellular filtrate [99]. The AuNP synthesis rate was found to increase with an increase in $HAuCl_4$ salt concentration, incubation time, and agitation, whereas pH and temperature showed a negative relation with the AuNP synthesis rate, which indicates higher productivity at lower values.

In the case of microbes, modification of pH leads to alteration in the overall charge of bioactive molecules, which in turn facilitates their binding affinity and hence biomineralization of metal ions into nanoparticles. For example, Yumei et al. showcased that *Arthrobacter* sp. promotes the synthesis of silver nanoparticles at pH 7.0 and 8.0, whereas no AgNP synthesis was recorded below pH 5.0 due to the strong electrostatic repulsion between silver ions and EPS in acidic conditions [73]. Even at higher pH (above 8), no AgNP formation was observed due to high electronegativity under alkaline conditions, which is not favorable for reduction of Ag^+ ions due to the presence of a $-COO^-$ group. Furthermore, process parameters could also affect the shape and yield of nanoparticles. Ramanathan et al. demonstrated that spherical-shaped silver nanoparticles (AgNPs) of small size were synthesized at an optimum growth temperature of 20°C using *Morganella psychrotolerans* [72], whereas silver nanoplates of large size were observed at a growth temperature of 4°C. Although there are a few reports on process parameter optimization, it is clear from the results of optimizing these variables could solve the issue of polydispersity and production yield of nanoparticles. This demands further investigation on the bio-mediated synthesis of nanoparticles for their production with high efficiency.

For efficient bio-synthesis of metallic nanoparticles, a number of controlling factors are involved in the nucleation and subsequent formation of stabilized nanoparticles. These factors include pH, reactant concentrations, reaction time, and temperature (as discussed earlier). Apart from optimizing these parameters, the use of biofilms is another prospective approach for efficient biosynthesis of nanoparticles [171]. Biofilms have recently been recognized as the most active growth mode of bacteria [172]. Biofilms exhibit various interesting properties such as catalyzing activity, highly reducing matrix, and the ability to control electrochemical reactions which provide a favorable environment for easy and efficient synthesis of nanoparticles than planktonic cells at forming nanoparticles [173]. Moreover, the protective nature of biofilms with diffusion limitation for outside materials, keeps the entire synthesis process free from contamination; overall, this makes it a promising approach for biosynthesis of nanoparticles in aqueous systems [174]. Nanoparticle synthesis in biofilms offers additional advantages, such as high biomass concentrations and large surface areas, which can lead to more efficient and scalable biosynthesis. Biofilms have up to 600 times higher metal resistance properties than their planktonic counterparts [175], and can catalyze electrochemical redox reactions by providing an appropriate environment with natural reducing agents such as proteins, peptides, and heterocyclic compounds for metal reduction to nanoparticles [174,176]. However, there has been very limited work on nanoparticle synthesis in biofilms and little is known about the stabilizing mechanism of nanoparticles in biofilms. Therefore, a thorough understanding of the molecular mechanism of nanoparticle synthesis in biofilms as well as their planktonic counterparts would help future researchers to develop more robust microbial systems for rapid and optimized biosynthesis of nanoparticles with desired sizes and shapes. Based on the mechanisms of metal reduction in bacterial biofilms, genetic modification of bacterial strains can be designed to obtain controlled sizes and shapes of nanoparticles and optimize production with high yield [177].

Furthermore, the conversion of metal into nanoparticles also brings toxicity issues. Several reports have mentioned adverse effects of these nanomaterials on biological systems and cellular components. The cytotoxicity of nanoparticles depends on various factors such as their size, shape, capping agent, density of nanoparticles, and the type of pathogens against which their toxicity is evaluated [178]. Nanoparticles synthesized from a nonbiological route are generally more toxic than those synthesized from the biological route. Some pathogens are more prone to nanoparticles, especially AgNPs, than others due to the presence of both the Ag^+ ions and NPs. They slowly envelope the microbial cell and enter inside it, inhibiting its essential metabolic functions. Nanoparticles are comparatively more toxic than bulk materials. They are toxic at cellular, subcellular, and molecular levels [179]. There are several reasons for the cytotoxicity of nanoparticles, such as physicochemical properties, contamination with toxic elements, small size, high surface charge, and free radical species generation. Oxidative stress and lipid peroxidation have been observed in fish brain tissue on exposure to nanoparticles [180]. Cytotoxicity by nanoparticles is thought to be generated through reactive oxygen species (ROS) as a result of which a decrease in glutathione levels and an increase in free radicals occur. Nanoparticles have a large surface area, which provides better contact with microbes. Therefore, these nanoparticles are able to penetrate the cell membrane or attach to the cell surface based on their particle size [181].

Moreover, they have been observed to be highly toxic to bacterial strains and their antibacterial efficacy is increased with a decrease in particle size. Carlson et al. have demonstrated an increase in ROS generation for 15-nm hydrocarbon-coated AgNPs as compared to 55 nm [182]. It has been observed by Liu et al. that 5-nm AgNPs were more toxic than 20- and 50-nm nanoparticles to four cell lines, namely, A549, HepG2, MCF-7, and SGC-7901 [183]. Till date, there has been extensive research into nanoparticle toxicity in order to explain their mechanism of action, and three different mechanisms have been devised so far which include cell wall and membrane damage, intracellular penetration and molecular damage, and oxidative stress. These modes of action are discussed in detail in the next paragraph.

The toxicity concern of nanoparticles can be suitably reduced by coating them with biocompatible agents. Although the main role of coating/capping is to stabilize nanoparticles and prevent agglomeration, their biocompatible nature also makes them suitable for various biomedical applications [178]. In green synthesis, stabilization of nanoparticles is achieved by the biocompatible material only, and hence the toxicity issue is reduced in most cases. In 2012, one report suggested that the stabilization of AgNPs by different polymer surfactants reduces the toxicity of AgNPs against mouse skin fibroblasts (L929), human hepatocarcinoma cells (HepG2), and mouse monocyte macrophages (J774A1) [184]. Polymer-capped AgNPs at a concentration of 1.5 ppm showed a hemocompatible nature. It is a well-known fact that the materials with a hemolysis ratio less than 5% are generally regarded as hemocompatible and safe [178,184]. Biosynthesis using polymer, i.e. glucan, resulted in glucan-capped spherical AgNPs of size 2.44 nm [185]. These nanoparticles showed only 0.68% hemolysis to human red blood cells (RBCs) at their LD50 dosage. Thus, AgNP–glucan conjugates were observed to be biocompatible with human RBCs at their LD50 dosage. In a similar study, polysaccharide-capped AgNPs of size 2.78 nm were synthesized using a hetero-polysaccharide isolated from *Lentinus squarrosulus* (Mont.) [186]. These nanoparticles also showed compatible nature with human RBCs at their LD50 dosage. Thus, the biocompatibility obtained through the green synthesis route suggests that it is possible to use nanoparticles in various biomedical applications.

9.5 CONCLUSION

In recent years, metal nanoparticles have been studied widely for various biomedical, bioremediation, and biosensor applications because of their remarkable antibacterial, antioxidant, and optical properties, large surface area to volume ratio, and higher efficacy. The synthesis of metal nanoparticles by biological mode has evolved as an important branch of nanobiotechnology, and bio-agents serve as potential nanofactories for the production of nanomaterials. However, there are some gaps and limitations in the successful production of these nanoparticles which need to be solved by the scientific community.

One of the major limitations in biomediated synthesis is a complete and thorough understanding of the mechanistic aspects of biofabrication of nanoparticles. Although there are reports in the literature on the identification and isolation of bioactive moiety responsible for biomineralization of metal ions using biological extracts, a much more detailed analysis of the biochemical pathway is needed for the development of tailor-made nanoparticles.

Especially for biomedical purposes, it is indispensable to understand how active moieties from various biological resources bind to the nanoparticle surface to provide stability, and to synthesize nanoparticles with higher biocompatibility. Large-scale production is another major bottleneck in the development and commercialization of biocompatible nanostructures with controlled sizes and shapes. Recently, researchers have focused on large-scale cultivation methods for nanoparticle synthesis which are scalable and reproducible with a narrow size distribution. However, these bulk cultivation methods for bionanomaterials and downstream processing techniques need to be improved further. Large-scale cultivation of nanoparticles is generally hampered by the factors of high cost, high energy requirement, polydispersity, and low nanoparticle yield. The production of nanoparticles at room temperature using natural active biomolecules without any reducing agents would make large-scale fermentation more cost-effective and energy sustainable. Stable production of monodispersed nanoparticles with high yield could be achieved by optimizing various process parameters (pH, temperature, contact time, mixing ratio, salt concentration) and altering the overall charge on functional molecules. Apart from this, issues related to the biomedical applications, namely the distribution profile, release kinetics, and clearance of nanostructures in vivo need to be addressed. In-depth evaluation of the biocompatibility and bioavailability of nanomaterials is still in its infancy and considerable research efforts are needed in this area. The collaborative research on fermentation process development along with an understanding of the mechanistic aspects, scale-up, and exploration of other biological agents could expedite the process of cost-effective tailor-made synthesis of nanomaterials.

REFERENCES

1. A. Arsha (2017). Bacterial synthesis and applications of nanoparticles. *Nano Sci. Nano Technol.*, 11, 119.
2. K. Alexander, S. Sheshrao Gajghate, A. Shankar Katarkar, A. Majumder and S. Bhaumik (2021). Role of nanomaterials and surfactants for the preparation of graphenenanofluid: A review. *Mater. Today: Proc.*, 44, 1136.
3. D. Cheng, Z. Qiao, L. Xuan and H.Wang (2020). Recent advances of morphology adaptive nanomaterials for anti-cancer drug delivery. *Prog.Nat. Sci-Mater*, 30, 555.
4. N. Rao, R. Singh and L. Bashambu (2021). Carbon-based nanomaterials: Synthesis and prospective applications. *Mater. Today: Proc.*, 44, 608.
5. S. Mazari, E. Ali, R. Abro, F. Khan, I. Ahmed, M. Ahmed, S. Nizamuddin, T. Siddiqui, N. Hossain, N. Mubarak and A. Shah (2021). Nanomaterials: Applications, waste-handling, environmental toxicities, and future challenges – A review. *J. Environ. Chem. Eng.*, 9, 105028.
6. Y. Wang, L. Wang, X. Zhang, X. Liang, Y. Feng and W. Feng (2021). Two-dimensional nanomaterials with engineered bandgap: Synthesis, properties, applications. *Nano Today*, 37, 101059.
7. A. Jawed, V. Saxena and L. Pandey (2020). Engineered nanomaterials and their surface functionalization for the removal of heavy metals: A review. *J. Water Process Eng.*, 33, 101009.

8. R. Tomar, A. Abdala, R. Chaudhary and N. Singh (2020). Photocatalytic degradation of dyes by nanomaterials. *Mater. Today: Proc.*, 29, 967–973.
9. K. Ai, J. Huang, Z. Xiao, Y. Yang and Y. Bai (2021). Localized surface plasmon resonance properties and biomedical applications of copper selenide nanomaterials. *Mater. Today Chem.*, 20, 100402.
10. J. Liu, D. Wu, N. Zhu, Y. Wu and G. Li (2021). Antibacterial mechanisms and applications of metal-organic frameworks and their derived nanomaterials. *Trend Food Sci. Tech.*, 109, 413–434.
11. J. Zhuang, C. He, K. Wang, K. Teng, Z. Ma, S. Zhang, L. Lu, X. Li, Y. Zhang and Q. An (2021). Nanoscopically-optimized carrier transportation and utilization in immobilized AuNP-TiO2 composite HER photocatalysts. *Appl. Surf. Sci.*, 537, 148055.
12. M. Nasrollahzadeh, S. Sajadi, M. Sajjadi and Z. Issaabadi (2019). "An Introduction to Nanotechnology," ch. 1 in *Interface Science and Technology*, Vol. 28, pp. 1–27. Amsterdam: Elsevier.
13. D. Schaming and H. Remita (2015). Nanotechnology: From the ancient time to nowadays. *Found. Chem.*, 17 (3), 187–205.
14. M. Nasrollahzadeh, S. M. Sajadi, M. Sajjadi and Z. Issaabadi (2019). *An Introduction to Green Nanotechnology*. San Diego: Elsevier.
15. H. Cheng, L. J. Doemeny, C. L. Geraci and D. Grob Schmidt (2016). *Nanotechnology Overview: Opportunities and Challenges*. Washington, DC: American Chemical Society.
16. F. Pastorino, C. Brignole, D. Di Paolo, P. Perri, F. Curnis, A. Corti and A., M. Ponzoni (2019). Overcoming biological barriers in neuroblastoma therapy: The vascular targeting approach with liposomal drug nanocarriers. *Small*, 15, 1804591. doi: 10.1002/smll.201804591
17. M. Sardar, A. Mishra and R. Ahmad (2014). Biosynthesis of metal nanoparticles and their applications," ch. 8, in *Biosensors and Nanotechnology*, eds. A. Tiwari and A. P. F. Turner, pp. 239–266. Beverly, MA: Scrivener. doi: 10.1002/9781118773826
18. R. Ahmad, S. Srivastava, S. Ghosh and S. K. Khare (2021). Phytochemical delivery through nanocarriers: A review. *Colloids Surf. B*, 197, 111389. doi: 10.1016/j.colsurfb.2020.111389
19. E. Blanco, H. Shen and M. Ferrari (2015). Principles of nanoparticle design for overcoming biological barriers to drug delivery. *Nat. Biotechnol.*, 33, 941. doi: 10.1038/nbt.3330
20. C. Gao, Y, Wang, Z. Ye, Z. Lin, X. Ma and Q. He (2020). Biomedical micro-/nanomotors: From overcoming biological barriers to in vivo imaging. *Adv. Mater.*, 33, 2000512. doi: 10.1002/adma.202000512
21. N. R. Stillman, M. Kovacevic, B. Igor and H. Sabine (2020). In silico modelling of cancer nanomedicine, across scales and transport barriers. *NPJ Comput. Mater.*, 6, 1–10. doi: 10.1038/s41524-020-00366-8
22. N. S. Aminabad, M. Farshbaf and A. Akbarzadeh (2019). Recent advances of gold nanoparticles in biomedical applications: State of the art. *Cell Biochem. Biophys.*, 77, 123–137. doi: 10.1007/s12013-018-0863-4
23. E. Boisselier and D. Astruc (2009). Gold nanoparticles in nanomedicine: Preparations, imaging, diagnostics, therapies and toxicity. *Chem. Soc. Rev.*, 38, 1759–1782. doi: 10.1039/b806051g
24. N. Elahi, M. Kamali and M. H. Baghersad (2019). Recent biomedical applications of gold nanoparticles: A review. *Talanta*, 184, 537–556. doi: 10.1016/j.talanta.2018.02.088
25. M. Chisanga, H. Muhamadali, D. I. Ellis and R. Goodacre (2019). Enhancing disease diagnosis: Biomedical applications of surface-enhanced Raman scattering. *Appl. Sci.*, 9, 1163. doi: 10.3390/app9061163

26. Y. Xia (2008). Nanomaterials at work in biomedical research. *Nat. Mater.,* 7, 758–760. doi: 10.1038/nmat2277

27. S. Celiksoy, W. Ye, K. Wandner, F. Schlapp, K. Kaefer, R. Ahijado-Guzman and C. Sönnichsen (2020). Plasmonic nanosensors for the label-free imaging of dynamic protein patterns. *J. Phys. Chem. Lett.,* 11, 4554–4558. doi: 10.1021/acs.jpclett.0c01400

28. R. M. P. Kumar, A. Venkatesh and V. H. S. Moorthy (2019). Nanopits based novel hybrid plasmonic nanosensor fabricated by a facile nanofabrication technique for biosensing. *Mater. Res. Express,* 6, 1150b6. doi: 10.1088/2053-1591/ab33b9

29. R. Noori, R. Ahmad and M. Sardar (2020). Nanobiosensor in health sector: the milestones achieved and future prospects, in *Nanobiosensors for Agricultural, Medical and Environmental Applications,* eds. M. Mohsin, R. Naz and A. Ahmad, pp. 63–90. Singapore: Springer. doi: 10.1007/978-981-15-8346-9_4

30. G. R. Rudramurthy and M. K. Swamy (2018). Potential applications of engineered nanoparticles in medicine and biology: An update. *J. Biol. Inorg. Chem.,* 23 (8), 1185–1204.

31. H. F. Florindo, R. Kleiner, D. Vaskovich-Koubi, R. C. Acurcio, B. Carreira, E. Yeini, G. Tiram, Y. Liubomirski and R. Satchi-Fainaro (2020). Immune-mediated approaches against COVID-19. *Nat. Nanotechnol.,* 15 (8), 630–645.

32. U. Sahin, A. Muik, E. Derhovanessian, I. Vogler, L. M. Kranz, M. Vormehr, et al. (2020). COVID-19 vaccine BNT162b1 elicits human antibody and TH1 T cell responses. *Nature,* 586, 594–599.

33. S. Talebian, G. G. Wallace, A. Schroeder, F. Stellacci and J. Conde (2020). Nanotechnology-based disinfectants and sensors for SARS-CoV-2. *Nat. Nanotechnol.,* 15 (8), 618–621.

34. S. Talebian and J. Conde Why go NANO on COVID-19 pandemic? *Matter,* 3 (3), 598–601.

35. P. N. Sudha, K. Sangeetha, K. Vijayalakshmi and A. Barhoum (2018). *Emerging Applications of Nanoparticles and Architecture Nanostructures.* Amsterdam: Elsevier.

36. F. Salamanca-Buentello, D. L. Persad, D. K. Martin, A. S. Daar and P. A. Singer (2005). Nanotechnology and the developing world. *PLoS Med.,* 2 (5), e97.

37. M. Nasrollahzadeh, S. M. Sajadi, M. Sajjadi and Z. Issaabadi (2019). *An Introduction to Green Nanotechnology.* San Diego: Elsevier.

38. StatNano.com. *Nanotechnology products.* https://product. statnano.com/ (accessed 12/ 06/2021).

39. T. Saleh and V. Gupta (2016). Synthesis of nanomaterial-incorporated membranes by physical methods, in *Nanomaterial and Polymer Membranes: Synthesis, Characterization, and Applications.* Elsevier: San Fransico, CA.

40. Z. Ma, P. Zhou, L. Zhang, Y. Zhong, X. Sui, B. Wang, Y. Ma, X. Feng, H. Xu, Z. Mao (2021). g-C_3N_4 nanosheets exfoliated by green wet ball milling process for photodegradation of organic pollutants. *Chem. Phys. Lett.,* 766, 138335.

41. T. Xie, L. Fu, W. Qin, J. Zhu, W. Yang, D. Li and L. Zhou (2018). Self-assembled metal nano-multilayered film prepared by co-sputtering method. *Appl. Surf. Sci.,* 435, 16.

42. T. Satyanarayana and S. Reddy (2018). A review on chemical and physical synthesis methods of nanomaterials. *IJASEIT.,* 6, 2321.

43. J. Li, S. Wang, X. Shi and M. Shen (2017). Aqueous-phase synthesis of iron oxide nanoparticles and composites for cancer diagnosis and therapy. *Adv. Colloid Interfac.,* 249, 374.

44. N. Hoda and F. Jamali-Sheini (2019). Influence of synthesis parameters on the physical properties of Cu3Se2 nanostructures using the sonochemical method. *Ceram. Int.,* 45, 16765.

45. L. Huang, J. Wang, Y. Zhu, Z. Li and K. Sun (2019). Effect of tio2–sio2 on microstructure and mechanical characteristics of zirconium corundum abrasives by sol-gel method. *J. Alloy. Compound.*, 802, 229.

46. M. Jalali-Jivan, F. Garavand and S. Jafari (2020). Microemulsions as nano-reactors for the solubilization, separation, purification and encapsulation of bioactive compounds. *Adv. Colloid. Interfac.*, 283, 102227.

47. Q. Sun, Y. Lv, X.Wu, W. Jia, J. Guo, F. Tong, D. Jia, Z. Sun and X. Wang (2020). Hydrothermal synthesis of coralloid-like vanadium nitride/carbon nanocomposites for high-performance and long-life supercapacitors. *J. Alloy. Compd.*, 818, 152895.

48. P. Zhang, D. Hou, X. Li, S. Pehkonen, R. Varma and X. Wang (2018). Green and size-specific synthesis of stable Fe–Cu oxides as earth-abundant adsorbents for malachite green removal. *ACS Sustain. Chem. Eng.*, 6, 9229.

49. A. Rana, K. Yadav and S. Jagadevan (2020). A comprehensive review on green synthesis of nature-inspired metal nanoparticles: Mechanism, application and toxicity. *J. Clean. Prod.*, 272, 122880.

50. S. Iravani (2011). Green synthesisofmetal nanoparticlesusingplants. *Green Chem.*, 13 (10), 2638–2650.

51. S. Iravani, H. Korbekandi, S. V. Mirmohammadi and H. Mekanik (2014). Plants in nanoparticle synthesis. *Rev. Adv. Sci. Eng.*, 3 (3), 261–274.

52. S. Iravani and B. Zolfaghari (2013). Green synthesis of silver nanoparticles using Pinus eldarica bark extract. *Biomed. Res. Int.,* 2013, 639725.

53. H. Korbekandi, Z.Ashari, S. Iravani and S.Abbasi (2013). Optimization of biological synthesis of silver nanoparticles using Fusarium oxysporum. *Iran. J. Pharm. Res.*, 12 (3), 289–298.

54. H. Korbekandi and S. Iravani (2013). Biological synthesis of nanoparticles using algae, in *Green Biosynthesis of Nanoparticles: Mechanisms and Applications*, eds. M. Rai and C. Posten, pp. 53–60. Wallingford, UK: CABI.

55. H. Korbekandi, S. Iravani and S. Abbasi (2009). Production of nanoparticles using organisms. *Crit. Rev. Biotechnol.*, 29 (4), 279–306.

56. H. Korbekandi, S. Iravani, and S. Abbasi (2012). Optimization of biological synthesis of silver nanoparticles using Lactobacillus casei subsp. casei. *J. Chem. Technol. Biotechnol.*, 87 (7), 932–937.

57. P. Singh, Y.-J. Kim, D. Zhang and D.-C. Yang (2016). Biological synthesis of nanoparticles from plants and microorganisms. *Trends Biotechnol.*, 34, 588–599.

58. S. Dhuper, D. Panda and P. Nayak (2012). Green synthesis and characterization of zero valent iron nanoparticles from the leaf extract of *Mangifera indica*. *Nano Trends: J. Nanotech. App.*, 13, 16–22.

59. A. Albanese, P. S. Tang and W. C. Chan (2012). The effect of nanoparticle size, shape, and surface chemistry on biological systems. *Annu. Rev. Biomed. Eng.*, 14, 1–16.

60. M. A. Faramarzi and A. Sadighi (2013). Insights into biogenic and chemical production of inorganic nanomaterials and nanostructures. *Adv. Colloid Interface Sci.*, 189, 1–20.

61. S. Iravani (2014). Bacteria in nanoparticle synthesis: Current status and future prospects. *Int. Sch. Res. Notices*, 2014, 1–18.

62. T. Beveridge and R. Murray (1980). Sites of metal deposition in the cell wall of *Bacillus subtilis*. *J. Bacteriol.*, 141, 876–887.

63. T. Klaus-Joerger, R. Joerger, E. Olsson and C.-G. Granqvist (2001). Bacteria as workers in the living factory: metal-accumulating bacteria and their potential for materials science. *Trends Biotechnol.*, 19, 15–20.

64. S. K. Srivastava and M. Constanti (2012). Room temperature biogenic synthesis of multiple nanoparticles (Ag, Pd, Fe, Rh, Ni, Ru, Pt, Co, and Li) by *Pseudomonas aeruginosa* SM1. *J. Nanopart. Res.*, 14, 831.

65. V. L. Das, R. Thomas, R. T. Varghese, E. Soniya, J. Mathew and E. Radhakrishnan (2014). Extracellular synthesis of silver nanoparticles by the *Bacillus* strain CS 11 isolated from industrialized area. *3 Biotech*, 4, 121–126.

66. R. R. Kulkarni, N. S. Shaiwale, D. N. Deobagkar and D. D. Deobagkar (2015). Synthesis and extracellular accumulation of silver nanoparticles by employing radiation-resistant *Deinococcus radiodurans*, their characterization, and determination of bioactivity. *Int. J. Nanomed.*, 10, 963.

67. E. Ahmed, S. Kalathil, L. Shi, O. Alharbi and P. Wang (2018). Synthesis of ultra-small platinum, palladium and gold nanoparticles by *Shewanella loihica* PV-4 electrochemically active biofilms and their enhanced catalytic activities. *J. Saudi Chem. Soc.*, 22, 919–929.

68. E. Zonaro, E. Piacenza, A. Presentato, F. Monti, R. Dell'Anna, S. Lampis and G. Vallini (2017). Ochrobactrum sp. MPV1 from a dump of roasted pyrites can be exploited as bacterial catalyst for the biogenesis of selenium and tellurium nanoparticles. *Microb. Cell Fact.*, 16, 215.

69. B. Srinath, K. Namratha and K. Byrappa (2018). Eco-friendly synthesis of gold nanoparticles by *Bacillus subtilis* and their environmental applications. *Adv. Sci. Lett.*, 24, 5942–5946.

70. M. Saravanan, S. K. Barik, D. Mubarak Ali, P. Prakash and A. Pugazhendhi (2018). Synthesis of silver nanoparticles from *Bacillus brevis* (NCIM 2533) and their antibacterial activity against pathogenic bacteria. *Microb. Pathog.*, 116, 221–226.

71. N. I. Hulkoti and T. Taranath (2014). Biosynthesis of nanoparticles using microbes – A review. *Colloids Surf. B*, 121, 474–483.

72. R. Ramanathan, A. P. O'Mullane, R. Y. Parikh, P. M. Smooker, S. K. Bhargava and V. Bansal (2010). Bacterial kinetics-controlled shape-directed biosynthesis of silver nanoplates using *Morganella psychrotolerans*. *Langmuir*, 27, 714–719.

73. L. Yumei, L. Yamei, L. Qiang and B. Jie (2017). Rapid biosynthesis of silver nanoparticles based on flocculation and reduction of an exopolysaccharide from *Arthrobacter* sp. B4: Its antimicrobial activity and phytotoxicity. *J. Nanomater.*, 2017, 1–8.

74. A. Malhotra, K. Dolma, N. Kaur, Y. S. Rathore, S. Mayilraj and A. R. Choudhury (2013). Biosynthesis of gold and silver nanoparticles using a novel marine strain of Stenotrophomonas. *Bioresour. Technol.*, 142, 727–731.

75. H. Kaur, K. Dolma, N. Kaur, A. Malhotra, N. Kumar, P. Dixit, D. Sharma, S. Mayilraj and A. R. Choudhury (2015). Biosynthesis of gold and silver nanoparticles using a novel marine strain of *Stenotrophomonas*. *Biotechnol. Bioprocess Eng.*, 20, 51–57.

76. G. Gahlawat, S. Shikha, B. S. Chaddha, S. R. Chaudhuri, S. Mayilraj and A. R. Choudhury (2016). Microbial glycolipoprotein-capped silver nanoparticles as emerging antibacterial agents against cholera. *Microb. Cell Fact.*, 15, 25.

77. A. Mehta, C. Sidhu, A. K. Pinnaka and A. R. Choudhury (2014). Extracellular polysaccharide production by a novel osmotolerant marine strain of *Alteromonas macleodii* its application towards biomineralization of silver. *PLoS One*, 9, e98798.

78. R. Raj, K. Dalei, J. Chakraborty and S. Das (2016). Extracellular polymeric substances of a marine bacterium mediated synthesis of CdS nanoparticles for removal of cadmium from aqueous solution. *J. Colloid Interface Sci.*, 462, 166–175.

79. S. Sunkar and C. V. Nachiyar (2012). Biogenesis of antibacterial silver nanoparticles using the endophytic bacterium *Bacillus cereus* isolated from *Garcinia xanthochymus*. *Asian Pac. J. Trop. Biomed.*, 2, 953–959.

80. M. Divya, S. K. George, H. Saqib and S. Joseph (2019). Biogenic synthesis and effect of silver nanoparticles (AgNPs) to combat catheter-related urinary tract infections. *Biocatal. Agric. Biotechnol.*, 18, 101037.
81. A.Müller, D. Behsnilian, E. Walz, V. Gräf, L. Hogekamp and R. Greiner (2016). Effect of culture medium on the extracellular synthesis of silver nanoparticles using *Klebsiella pneumoniae, Escherichia coli* and *Pseudomonas jessinii. Biocatal. Agric. Biotechnol.*, 6, 107–115.
82. M. M. Naik, M. S. Prabhu, S. N. Samant, P. M. Naik and S. Shirodkar (2017). Siderophore production by bacteria isolated from mangrove sediments: A microcosm study. *Thalassas: Int. J. Mar. Sci.*, 33, 73–80.
83. M. Manimaran and K. Kannabiran (2017). Actinomycetes-mediated biogenic synthesis of metal and metal oxide nanoparticles: Progress and challenges. *Lett. Appl. Microbiol.*, 64, 401–408.
84. S. Otari, R. Patil, N. Nadaf, S. Ghosh and S. Pawar (2012). Green biosynthesis of silver nanoparticles from an actinobacteria *Rhodococcus* sp. *Mater. Lett.*, 72, 92–94.
85. L. Karthik, G. Kumar, A. V. Kirthi, A. Rahuman and K. B. Rao (2014). *Streptomyces* sp. LK3 mediated synthesis of silver nanoparticles and its biomedical application. *Bioprocess Biosyst. Eng.*, 37, 261–267.
86. B. Buszewski, V. Railean-Plugaru, P. Pomastowski, K. Rafinska, M. Szultka-Mlynska, P. Golinska, M. Wypij, D. Laskowski and H. Dahm (2016). Antimicrobial activity of biosilver nanoparticles produced by a novel *Streptacidiphilusdurhamensis* strain. *J. Microbiol. Immunol. Infect.*, 20, 1–10.
87. H. M. Abd-Elnaby, G. M. Abo-Elala, U. M. Abdel-Raouf and M. M. Hamed (2016). Antibacterial and anticancer activity of extracellular synthesized silver nanoparticles from marine *Streptomyces rochei* MHM13. *Egypt. J. Aquat. Res.*, 42, 301–312.
88. V. Ranjitha and V. R. Rai (2017). Antibacterial and anticancer activity of extracellular synthesized silver nanoparticles from marine *Streptomyces rochei* MHM13. *3 Biotech*, 7, 299.
89. E. Saad, S. S. Salem, A. Fouda, M. A. Awad, M. S. El-Gamal and A. M. Abdo (2018). New approach for antimicrobial activity and bio-control of various pathogens by biosynthesized copper nanoparticles using endophytic actinomycetes. *J. Radiat. Res. Appl. Sci.*, 30, 1–9.
90. S. Otari, R. Patil, N. Nadaf, S. Ghosh and S. Pawar (2014). Green synthesis of silver nanoparticles by microorganism using organic pollutant: its antimicrobial and catalytic application. *Environ. Sci. Pollut. Res.*, 21, 1503–1513.
91. N. Mara Silva-Vinhote, N. E. D. Caballero, T. de Amorim Silva, P. V. Quelemes, A. R. de Araujo, A. C. M. de Moraes, A. L. dos Santos Cˆamara, J. P. F. Longo, R. B. Azevedo and D. A. da Silva (2017). Extracellular biogenic synthesis of silver nanoparticles by Actinomycetes from amazonic biome and its antimicrobial efficiency. *Afr. J. Biotechnol.*, 16, 2072–2082.
92. M. Wypij, J. Czarnecka, M. Świecimska, H. Dahm, M. Rai and P. Golinska (2018). Synthesis, characterization and evaluation of antimicrobial and cytotoxic activities of biogenic silver nanoparticles synthesized from *Streptomyces xinghaiensis* OF1 strain. *World J. Microbiol. Biotechnol.*, 34, 23.
93. A. Bhargava, N. Jain, M. A. Khan, V. Pareek, R. V. Dilip and J. Panwar (2016). Utilizing metal tolerance potential of soil fungus for efficient synthesis of gold nanoparticles with superior catalytic activity for degradation of rhodamine B. *J. Environ. Manage.*, 183, 22–32.
94. A. Mishra, M. Kumari, S. Pandey, V. Chaudhry, K. Gupta and C. Nautiyal (2014). Biocatalytic and antimicrobial activities of gold nanoparticles synthesized by *Trichoderma* sp. *Bioresour. Technol.*, 166, 235–242.

95. R. P. Metuku, S. Pabba, S. Burra, K. Gudikandula and M. S. Charya (2014). Biosynthesis of silver nanoparticles from *Schizophyllum radiatum* HE 863742.1: Their characterization and antimicrobial activity. *3 Biotech*, 4, 227–234.
96. S. Rajput, R. Werezuk, R. M. Lange and M. T. McDermott (2016). Fungal isolate optimized for biogenesis of silver nanoparticles with enhanced colloidal stability. *Langmuir*, 32, 8688–8697.
97. M. Kitching, P. Choudhary, S. Inguva, Y. Guo, M. Ramani, S. K. Das and E. Marsili (2016). Fungal surface protein mediated one-pot synthesis of stable and hemocompatible gold nanoparticles. *Enzyme Microb. Technol.*, 95, 76–84.
98. P. Suryavanshi, R. Pandit, A. Gade, M. Derita, S. Zachino and M. Rai (2017). *Colletotrichum* sp.-mediated synthesis of sulphur and aluminium oxide nanoparticles and its in vitro activity against selected food-borne pathogens. *LWT–Food Sci. Technol.*, 81, 188–194.
99. E. B. El Domany, T. M. Essam, A. E. Ahmed and A. A. Farghali (2018). Biosynthesis physico-chemical optimization of gold nanoparticles as anti-cancer and synergetic antimicrobial activity using *Pleurotus ostreatus* fungus. *J. Appl. Pharm. Sci.*, 8, 119–128.
100. M. Hamad (2018). Removal of phenol and inorganic metals from wastewater using activated ceramic. *Int. J. Environ. Sci. Technol.*, 1–10.
101. S. V. Ganachari, R. Bhat, R. Deshpande and A. Venkataraman (2012). Extracellular biosynthesis of silver nanoparticles using fungi *Penicillium diversum* and their anti-microbial activity studies. *BioNanoScience*, 2, 316–321.
102. S. Roy, T. Mukherjee, S. Chakraborty and T. K. Das (2013). Biosynthesis, characterisation & antifungal activity of silver nanoparticles synthesized by the fungus *Aspergillus foetidus* MTCC8876. *Dig. J. Nanomater. Biostruct.*, 8, 197–205.
103. J. N. Thakker, P. Dalwadi and P. C. Dhandhukia (2012). Biosynthesis of gold nanoparticles using *Fusarium oxysporum* f. sp. *cubense* JT1, a plant pathogenic fungus. *ISRN Biotechnol.*, 2013, 515091.
104. A. S. Bhadwal, R. Tripathi, R. K. Gupta, N. Kumar, R. Singh and A. Shrivastav (2014). Biogenic synthesis and photocatalytic activity of CdS nanoparticles. *RSC Adv.*, 4, 9484–9490.
105. M. Castro, L. Cottet and A. Castillo (2014). Biosynthesis of gold nanoparticles by extracellular molecules produced by the phytopathogenic fungus Botrytis cinerea. *Mater. Lett.*, 115, 42–44.
106. P. K. Kar, S. Murmu, S. Saha, V. Tandon and K. Acharya (2014). Anthelmintic efficacy of gold nanoparticles derived from a phytopathogenic fungus, *Nigrospora oryzae*. *PLoS One*, 9, e84693.
107. G. Baskar, J. Chandhuru, K. S. Fahad, A. Praveen, M. Chamundeeswari and T. Muthukumar (2015). Anticancer activity of fungal L-asparaginase conjugated with zinc oxide nanoparticles. *J. Mater. Sci. Mater. Med.*, 26, 43.
108. P. Ramalingmam, S. Muthukrishnan and P. Thangaraj (2015). Biosynthesis of silver nanoparticles using an endophytic fungus, *Curvularia lunata* and its antimicrobial potential. *J. Nanosci. Nanoeng.*, 1, 241–247.
109. D. Amerasan, T. Nataraj, K. Murugan, C. Panneerselvam, P. Madhiyazhagan, M. Nicoletti and G. Benelli (2016). Myco-synthesis of silver nanoparticles using Metarhizium anisopliae against the rural malaria vector *Anopheles culicifacies* Giles (Diptera: Culicidae). *J. Pest Sci.*, 89, 249–256.
110. M. Guilger, T. Pasquoto-Stigliani, N. Bilesky-Jose, R. Grillo, P. Abhilash, L. F. Fraceto and R. De Lima (2017). Biogenic silver nanoparticles based on *Trichoderma*

harzianum: Synthesis, characterization, toxicity evaluation and biological activity. *Sci. Rep.*, 7, 44421.

111. S. Hamedi, M. Ghaseminezhad, S. Shokrollahzadeh and S. A. Shojaosadati (2017). Controlled biosynthesis of silver nanoparticles using nitrate reductase enzyme induction of filamentous fungus and their antibacterial evaluation. *Artif. Cells Nanomed. Biotechnol.*, 45, 1588–1596.

112. M. Shah, D. Fawcett, S. Sharma, S. Tripathy and G. Poinern (2015). Green synthesis of metallic nanoparticles via biological entities. *Materials* 8, 7278–7308.

113. M. Apte, D. Sambre, S. Gaikawad, S. Joshi, A. Bankar, A. R. Kumar and S. Zinjarde (2013). Psychrotrophic yeast *Yarrowia lipolytica* NCYC 789 mediates the synthesis of antimicrobial silver nanoparticles via cell-associated melanin. *AMB Express*, 3, 32.

114. S. R. Waghmare, M. N. Mulla, S. R. Marathe and K. D. Sonawane (2015). Ecofriendly production of silver nanoparticles using *Candida utilis* and its mechanistic action against pathogenic microorganisms. *3 Biotech*, 5, 33–38.

115. F. Elahian, S. Reiisi, A. Shahidi and S. A. Mirzaei (2017). High-throughput bioaccumulation, biotransformation, and production of silver and selenium nanoparticles using genetically engineered *Pichia pastoris*. *Nanomedicine,* 13, 853–861.

116. M. Eugenio, N. M¨uller, S. Fras´es, R. Almeida-Paes, L. M. T. Lima, L. Lemgruber, M. Farina, W. de Souza and C. Sant'Anna (2016). Yeast-derived biosynthesis of silver/ silver chloride nanoparticles and their antiproliferative activity against bacteria. *RSC Adv.*, 6, 9893–9904.

117. M. Sriramulu and S. Sumathi (2018). Biosynthesis of palladium nanoparticles using *Saccharomyces cerevisiae* extract and its photocatalytic degradation behaviour. *Adv. Nat. Sci. Nanosci. Nanotechnol.*, 9, 025018.

118. S. Seshadri, K. Saranya and M. Kowshik (2011). Green synthesis of lead sulfide nanoparticles by the lead resistant marine yeast, *Rhodosporidium diobovatum*. *Biotechnol. Prog.*, 27, 1464–1469.

119. F. G. Ortega, M. A. Fern´andez-Baldo, J. G. Fern´andez, M. J. Serrano, M. I. Sanz, J. J. D´ıaz-Moch´on, J. A. Lorente and J. Raba (2015). Study of antitumor activity in breast cell lines using silver nanoparticles produced by yeast. *Int. J. Nanomed.*, 10, 2021.

120. H. Korbekandi, S. Mohseni, R. Mardani Jouneghani, M. Pourhossein and S. Iravani (2016). Biosynthesis of silver nanoparticles using *Saccharomyces cerevisiae*. *Artif. Cells Nanomed. Biotechnol.*, 44, 235–239.

121. X. Zhang, Y. Qu, W. Shen, J. Wang, H. Li, Z. Zhang, S. Li and J. Zhou (2016). Biogenic synthesis of gold nanoparticles by yeast *Magnusiomyces ingens* LH-F1 for catalytic reduction of nitrophenols. *Colloids Surf. A*, 497, 280–285.

122. J. G. Fern´andez, M. A. Fern´andez-Baldo, E. Berni, G. Cam´ı, N. Dur´an, J. Raba and M. I. Sanz (2016). Production of silver nanoparticles using yeasts and evaluation of their antifungal activity against phytopathogenic fungi. *Process Biochem.*, 51, 1306–1313.

123. Z. Yang, Z. Li, X. Lu, F. He, X. Zhu, Y. Ma, R. He, F. Gao, W. Ni and Y. Yi (2017). Controllable biosynthesis and properties of gold nanoplates using yeast extract. *Nano-Micro Lett.*, 9, 5.

124. J. J. A. Bonilla, D. J. P. Guerrero, R. G. T. Śaez, K. Ishida, B. B. Fonseca, S. Rozental and C. C. O. Ĺopez (2017). Green synthesis of silver nanoparticles using maltose and cysteine and their effect on cell wall envelope shapes and microbial growth of *Candida* spp. *J. Nanosci. Nanotechnol.*, 17, 1729–1739.

125. A. B. Moghaddam, M. Moniri, S. Azizi, R. A. Rahim, A. B. Ariff, W. Z. Saad, F. Namvar, M. Navaderi and R. Mohamad (2017). Biosynthesis of ZnO nanoparticles by

a new *Pichia kudriavzevii* yeast strain and evaluation of their antimicrobial and anti-oxidant activities. *Molecules, 22*, 872.

126. M. R. Salvadori, R. A. Ando, C. A. O. Nascimento and B. Corrêa (2017). Dead bio-mass of Amazon yeast: A new insight into bioremediation and recovery of silver by intracellular synthesis of nanoparticles. *J. Environ. Sci. Health Part A*, 52, 1112–1120.

127. A. Rónavári, N. Igaz, M. K. Gopisetty, B. Szerencśes, D. Kovacs, C. Papp, C. Vagvolgyi, I. M. Boros, Z. Kónya and M. Kiricsi (2018). Biosynthesized silver and gold nanoparticles are potent antimycotics against opportunistic pathogenic yeasts and dermatophytes. *Int. J. Nanomed.*, 13, 695.

128. Y. Qu, S. You, X. Zhang, X. Pei, W. Shen, Z. Li, S. Li and Z. Zhang (2018). Biosynthesis of gold nanoparticles using cell-free extracts of *Magnusiomyces ingens* LH-F1 for nitrophenols reduction. *Bioprocess Biosyst. Eng.*, 41, 359–367.

129. F. A. Cunha, M. da CSO Cunha, S. M. da Frota, E. J. Mallmann, T. M. Freire, L. S. Costa, A. J. Paula, E. A. Menezes and P. B. Fechine (2018). Biogenic synthesis of multifunctional silver nanoparticles from *Rhodotorula glutinis and Rhodotorula mucilaginosa*: antifungal, catalytic and cytotoxicity activities. *World J. Microbiol. Biotechnol.*, 34, 127.

130. M. Jalal, M. Ansari, M. Alzohairy, S. Ali, H. Khan, A. Almatroudi and K. Raees (2018). Biosynthesis of silver nanoparticles from oropharyngeal *Candida glabrata* isolates and their antimicrobial activity against clinical strains of bacteria and fungi. *Nanomaterials*, 8, 586.

131. V. da Silva Ferreira, M. E. ConzFerreira, L. M. T. Lima, S. Fraśes, W. de Souza and C. Sant'Anna (2017). Green production of microalgae-based silver chloride nanoparticles with antimicrobial activity against pathogenic bacteria. *Enzyme Microb. Technol.*, 97, 114–121.

132. F. Arsiya, M. H. Sayadi and S. Sobhani (2017). Green synthesis of palladium nanoparticles using *Chlorella vulgaris*. *Mater. Lett.*, 186, 113–115.

133. S. Momeni and I. Nabipour (2015). A simple green synthesis of palladium nanoparticles with *Sargassum* alga and their electrocatalytic activities towards hydrogen peroxide *Appl. Biochem. Biotechnol.*, 176, 1937–1949.

134. T. S. Dhas, V. G. Kumar, V. Karthick, K. J. Angel and K. Govindaraju (2014). Facile synthesis of silver chloride nanoparticles using marine alga and its antibacterial efficacy. *Spectrochim. Acta Part A*, 120, 416–420.

135. T. N. J. I. Edison, R. Atchudan, C. Kamal and Y. R. Lee (2016). *Caulerpa racemosa*: A marine green alga for eco-friendly synthesis of silver nanoparticles and its catalytic degradation of methylene blue. *Bioprocess Biosyst. Eng.*, 39, 1401–1408.

136. M. Ramakrishna, D. R. Babu, R. M. Gengan, S. Chandra and G. N. Rao (2016). Green synthesis of gold nanoparticles using marine algae and evaluation of their catalytic activity. *J. Nanostruct. Chem.*, 6, 1–13.

137. S. Rajeshkumar (2018). Synthesis of zinc oxide nanoparticles using algal formulation (*Padina tetrastromatica and Turbinaria conoides*) and their antibacterial activity against fish pathogens. *Res. J. Biotechnol.*, 13, 15–19.

138. Z. Sanaeimehr, I. Javadi and F. Namvar (2018). Antiangiogenic and antiapoptotic effects of green-synthesized zinc oxide nanoparticles using *Sargassum muticum* algae extraction. *Cancer Nanotechnol.*, 9, 3.

139. D. Fawcett, J. J. Verduin, M. Shah, S. B. Sharma and G. E. J. Poinern (2017). A review of current research into the biogenic synthesis of metal and metal oxide nanoparticles via marine algae and seagrasses. *J. Nanosci.*, 2017, 1–15.

140. V. Venkatpurwar and V. Pokharkar (2011). Green synthesis of silver nanoparticles using marine polysaccharide: Study of *in-vitro* antibacterial activity. *Mater. Lett.*, 65, 999–1002.

141. F. A. A. Rajathi, C. Parthiban, V. G. Kumar and P. Anantharaman (2012). Biosynthesis of antibacterial gold nanoparticles using brown alga, *Stoechospermum marginatum* (kützing). *Spectrochim. Acta, Part A*, 99, 166–173.

142. S. Senapati, A. Syed, S. Moeez, A. Kumar and A. Ahmad (2012). Intracellular synthesis of gold nanoparticles using alga *Tetraselmis kochinensis*. *Mater. Lett.*, 79, 116–118.

143. R. R. R. Kannan, R. Arumugam, D. Ramya, K. Manivannan and P. Anantharaman (2013). Green synthesis of silver nanoparticles using marine macroalga *Chaetomorpha linum*. *Appl. Nanosci.*, 3, 229–233.

144. Z. Salari, F. Danafar, S. Dabaghi and S. A. Ataei (2016). Sustainable synthesis of silver nanoparticles using macroalgae *Spirogyra varians* and analysis of their antibacterial activity. *J. Saudi Chem. Soc.*, 20, 459–464.

145. J. Jena, N. Pradhan, R. R. Nayak, B. P. Dash, L. B. Sukla, P. K. Panda and B. K. Mishra (2014). Microalga *Scenedesmus* sp.: A potential low-cost green machine for silver nanoparticle synthesis. *J. Microbiol. Biotechnol.*, 24, 522–533.

146. J. Annamalai and T. Nallamuthu (2015). Characterization of biosynthesized gold nanoparticles from aqueous extract of *Chlorella vulgaris* and their anti-pathogenic properties. *Appl. Nanosci.*, 5, 603–607.

147. J. Venkatesan, P. Manivasagan, S.-K. Kim, A. V. Kirthi, S. Marimuthu and A. A. Rahuman (2014). Marine algae-mediated synthesis of gold nanoparticles using a novel *Ecklonia cava*. *Bioprocess Biosyst. Eng.*, 37, 1591–1597.

148. T. Kathiraven, A. Sundaramanickam, N. Shanmugam and T. Balasubramanian (2015). Green synthesis of silver nanoparticles using marine algae *Caulerpa racemosa* and their antibacterial activity against some human pathogens. *Appl. Nanosci.*, 5, 499–504.

149. K. Murugan, C. M. Samidoss, C. Panneerselvam, A. Higuchi, M. Roni, U. Suresh, B. Chandramohan, J. Subramaniam, P. Madhiyazhagan and D. Dinesh (2015). Seaweed-synthesized silver nanoparticles: An eco-friendly tool in the fight against *Plasmodium falciparum* and its vector *Anopheles stephensi*? *Parasitol. Res.*, 114, 4087–4097.

150. S. N. Sinha, D. Paul, N. Halder, D. Sengupta and S. K. Patra (2015). Green synthesis of silver nanoparticles using fresh water green alga *Pithophora oedogonia* (Mont.) Wittrock and evaluation of their antibacterial activity. *Appl. Nanosci.*, 5, 703–709.

151. N. Gonźalez-Ballesteros, S. Prado-López, J. Rodriguez-Gonzalez, M. Lastra and M. Rodríguez-Argüelles (2017). Green synthesis of gold nanoparticles using brown algae *Cystoseira baccata*: Its activity in colon cancer cells. *Colloids Surf. B*, 153, 190–198.

152. N. Abdel-Raouf, N. M. Al-Enazi and I. B. Ibraheem (2017). Green biosynthesis of gold nanoparticles using *Galaxaura elongata* and characterization of their antibacterial activity. *Arabian J. Chem.*, 10, S3029–S3039.

153. D.-Y. Kim, R. G. Saratale, S. Shinde, A. Syed, F. Ameen and G. Ghodake (2018). Green synthesis of silver nanoparticles using *Laminaria japonica* extract: Characterization and seedling growth assessment. *J. Cleaner Prod.*, 172, 2910–2918.

154. A. Pugazhendhi, D. Prabakar, J. M. Jacob, I. Karuppusamy and R. G. Saratale (2018). Synthesis and characterization of silver nanoparticles using *Gelidium amansii* and its antimicrobial property against various pathogenic bacteria. *Microb. Pathog.*, 114, 41–45.

155. M. Kobayashi, S. Tomita, K. Sawada, K. Shiba, H. Yanagi, I. Yamashita and Y. Uraoka (2012). Chiral meta-molecules consisting of gold nanoparticles and genetically engineered tobacco mosaic virus. *Opt. Express*, 20, 24856–24863.

156. Q. Zeng, H. Wen, Q. Wen, X. Chen, Y. Wang, W. Xuan, J. Liang and S. Wan (2013). Cucumber mosaic virus as drug delivery vehicle for doxorubicin. *Biomaterials*, 34, 4632–4642.

157. C. Mao, C. E. Flynn, A. Hayhurst, R. Sweeney, J. Qi, G. Georgiou, B. Iverson and A. M. Belcher (2003). Viral assembly of oriented quantum dot nanowires. *Proc. Natl. Acad. Sci. U. S. A.*, 100, 6946–6951.

158. A. J. Love, M. E. Talianski, S. N. Chapman and J. Shaw (2017). A review on the biosynthesis of metal and metal salt nanoparticles by microbes. US Pat., No. 9,688,964, U.S. Patent and Trademark Office, Washington, DC, 2017.

159. J. Cao, R. H. Guenther, T. L. Sit, C. H. Opperman, S. A. Lommel and J. A. Willoughby (2014). Loading and release mechanism of red clover necrotic mosaic virus derived plant viral nanoparticles for drug delivery of doxorubicin. *Small,* 10, 5126–5136.

160. D. H. Le, K. L. Lee, S. Shukla, U. Commandeur and N. F. Steinmetz (2017). Potato virus X, a filamentous plant viral nanoparticle for doxorubicin delivery in cancer therapy. *Nanoscale*, 9, 2348–2357.

161. F. Yang, Y. Li, T. Liu, K. Xu, L. Zhang, C. Xu and J. Gao (2013). Plasma synthesis of Pd nanoparticles decorated-carbon nanotubes and its application in Suzuki reaction. *Chem. Eng. J.,* 226, 52–58.

162. P.-Y. Chen, X. Dang, M. T. Klug, N.-M. D. Courchesne, J. Qi, M. N. Hyder, A. M. Belcher and P. T. Hammond (2015). M13 virus-enabled synthesis of titanium dioxide nanowires for tunable mesoporous semiconducting networks. *Chem. Mater.*, 27, 1531–1540.

163. N. Esfandiari, M. K. Arzanani, M. Soleimani, M. Kohi- Habibi and W. E. Svendsen (2016). A new application of plant virus nanoparticles as drug delivery in breast cancer. *Tumor Biol.*, 37, 1229–1236.

164. C. C. Chen, M. Stark, M. Baikoghli and R. H. Cheng (2018). Surface functionalization of hepatitis E virus nanoparticles using chemical conjugation methods. *J. Visualized Exp.*, e57020.

165. J.-W. Moon, I. N. Ivanov, P. C. Joshi, B. L. Armstrong, W. Wang, H. Jung, A. J. Rondinone, G. E. Jellison Jr, H. M. Meyer III and G. G. Jang (2014). Scalable production of microbially mediated zinc sulfide nanoparticles and application to functional thin films. *Acta Biomater.,* 10, 4474–4483.

166. A. Ramos-Ruiz, J. Sesma-Martin, R. Sierra-Alvarez and J. A. Field (2017). Continuous reduction of tellurite to recoverable tellurium nanoparticles using an upflow anaerobic sludge bed (UASB) reactor. *Water Res.,* 108, 189–196.

167. J.-W. Moon, C. J. Rawn, A. J. Rondinone, L. J. Love, Y. Roh, S. M. Everett, R. J. Lauf and T. J. Phelps (2010). Large-scale production of magnetic nanoparticles using bacterial fermentation. *J. Ind. Microbiol. Biotechnol.*, 37, 1023–1031.

168. J.-W. Moon, I. N. Ivanov, C. E. Duty, L. J. Love, A. J. Rondinone, W. Wang, Y.-L. Li, A. S. Madden, J. J. Mosher and M. Z. Hu (2013). Article navigation scalable economic extracellular synthesis of CdS nanostructured particles by a non-pathogenic thermophile. *J. Ind. Microbiol. Biotechnol.*, 40, 1263–1271.

169. J.-W. Moon, T. J. Phelps, C. L. Fitzgerald Jr, R. F. Lind, J. G. Elkins, G. G. Jang, P. C. Joshi, M. Kidder, B. L. Armstrong and T. R. Watkins (2016). Manufacturing demonstration of microbially mediated zinc sulfide nanoparticles in pilot-plant scale reactors. *Appl. Microbiol. Biotechnol.*, 100, 7921–7931.

170. S. Ahmed, M. Ahmad, B. L. Swami and S. Ikram (2016). A review on plants extract mediated synthesis of silver nanoparticles for antimicrobial applications: A green expertise. *J. Adv. Res.*, 7, 17–28

171. A. H. Tanzil, S. T. Sultana, S. R. Saunders, L. Shi, E. Marsili and H. Beyenal (2016). Biological synthesis of nanoparticles in biofilms. *Enzyme Microb. Technol.*, 95, 4–12.

172. K. Ikuma, A. W. Decho and B. L. T. Lau (2015). When nanoparticles meet biofilms – Interactions guiding the environmental fate and accumulation of nanoparticles. *Front. Microbiol.*, 6, 1–6.

173. S. Kalathil, J. Lee and M. H. Cho (2011). Electrochemically active biofilm-mediated synthesis of silver nanoparticles in water. *Green Chem.*, 13, 1482–1485.

174. M. M. Khan, S. Kalathil, T. H. Han, J. Lee and M. H. J. Cho (2013). Positively charged gold nanoparticles synthesized by electrochemically active biofilm – A biogenic approach. *J. Nanosci. Nanotechnol.*, 13, 6079–6085.

175. G. M. Teitzel and M. R. Parsek (2003). Heavy metal resistance of biofilm and planktonic *Pseudomonas aeruginosa. Appl. Environ. Microbiol.*, 69, 2313–2320.

176. M. M. Khan, S. A. Ansari, J. H. Lee, M. O. Ansari, J. Lee and M. H. Cho (2014). Electrochemically active biofilm assisted synthesis of Ag@CeO2 nanocomposites for antimicrobial activity, photocatalysis and photoelectrodes. *J. Colloid Interface Sci.*, 431, 255–263.

177. D. Mandal, M. E. Bolander, D. Mukhopadhyay, G. Sarkar and P. Mukherjee (2006). The use of microorganisms for the formation of metal nanoparticles and their application. *Appl. Microbiol. Biotechnol.,* 69, 485–492.

178. A. Roy, O. Bulut, S. Some, A. K. Mandal and M. D. Yilmaz (2019). Green synthesis of silver nanoparticles: Biomolecule-nanoparticle organizations targeting antimicrobial activity. *RSC Adv.*, 9, 2673–2702.

179. L. Jayasree, P. Janakiram and R. Madhavi (2006). Characterization of Vibrio spp. associated with diseased shrimp from culture ponds of Andhra Pradesh (India). *J. World Aquacult. Soc.,* 37, 523–532.

180. E. Oberdorster (2004). Manufactured nanomaterials (fullerenes, C60) induce oxidative stress in the brain of juvenile largemouth bass. *Environ. Health Perspect.,* 112, 1058–1062.

181. M. Rai, A. Yadav and A. Gade (2009). Silver nanoparticles as a new generation of antimicrobials. *Biotechnol. Adv.*, 27, 76–83.

182. C. Carlson, S. M. Hussain, A. M. Schrand, L. K. Braydich-Stolle, K. L. Hess, R. L. Jones and J. J. Schlager (2008). Unique cellular interaction of silver nanoparticles: Size-dependent generation of reactive oxygen species. *J. Phys. Chem. B*, 112, 13608–13619.

183. W. Liu, Y. Wu, C. Wang, H. C. Li, T. Wang, C. Y. Liao, L. Cui, Q. F. Zhou, B. Yan and G. B. Jiang (2010). Impact of silver nanoparticles on human cells: Effect of particle size. *Nanotoxicology*, 4, 319–330.

184. J. J. Lin, W. C. Lin, R. X. Dong and S. Hsu (2012). The cellular responses and antibacterial activities of silver nanoparticles stabilized by different polymers. *Nanotechnology,* 23, 065102.

185. I. K. Sen, A. K. Mandal, S. Chakraborti, B. Dey, R. Chakraborty and S. S. Islam (2013). Green synthesis of silver nanoparticles using glucan from mushroom and study of antibacterial activity. *Int. J. Biol. Macromol.*, 62, 439–449.

186. D. K. Manna, A. K. Mandal, I. K. Sen, P. K. Maji, S. Chakraborti, R. Chakraborty and S. S. Islam (2015). Antibacterial and DNA degradation potential of silver nanoparticles synthesized via green route. *Int. J. Biol. Macromol.,* 80, 455–459.

10 Nanoparticles

Applications in Pharmaceutical Sciences, Cancer Diagnosis, and Therapeutics

Anand Pithadia, Bhavisha Patel, Anil Jogdand, and Vijay J. Upadhye

10.1 NANOPARTICLES IN THE DEVELOPMENT OF HERBALS AND HERBAL FORMULATIONS

Nanoparticles are of enormous scientific significance because they bridge the gap between atomic or molecular structures, and are concerned with materials at the nanolevel [1–6]. Nanomaterials exhibit unique physical, chemical, and biological attributes due to their small size, shape, composition, distribution, and morphology, in contrast to micro and macro materials. Nanoparticles are described as those particulate materials having at least one dimension ranging between 1–100 nanometers (nm). The term "nano" was derived from the Greek word for "dwarf." The famous physicist, Richard Feynman realized the concept of nanotechnology in 1959 in his pioneering lecture "There's plenty of room at the bottom," organized by the American Physical Society, where he cited the prospect of modifying material at atomic and molecular levels. With the advent of the 21st century, modern medicinal science has witnessed rapid and innovative progress worldwide in the field of research at the nano-scale [7]. However, the use of nanoparticles in the form of "Bhasma" was first described more than 5,000 years ago in Ayurveda, the Traditional Indian System of Medicine. Traditional methods of healing usually employ herbs as an integral part of their practice all over the world.

Although a number of nanoparticle-based products have great prospects as potential drug leads, in addition to these nanoparticle-based products, plant-based products are also reclaiming their past glory in the field of pharmaceutical development, since the drug targets are diminishing and combinatorial synthetic methods are not satisfactory. The uniqueness of the present thesis lies in the study of the synergistic effects of nanoparticles and plant products. Plants have always been an exemplary source of potent medicines. The majority of the existing drugs have been obtained from plants,

DOI: 10.1201/9781003362258-10

either directly or indirectly. In the Indian medicinal system, many physicians prepare and practice their own formulae, which require thorough scientific validation.

The World Health Organization predicted that 80% of the global population depends on herbal medicines [8]. The usage of herbal medicines is gradually increasing for curing medical disorders in the Western population, and the United States is also moving towards natural or organic therapies, enhancing thereby the use of herb-based medicines. German practitioners have recommended nearly 700 phytomedicines [9]. Asian countries like India and China are dependent upon conventional modes of medication.

The growing occurrence of undesirable side effects and the higher cost of existing drugs have led to the interest of scientists in conventional drugs as a complementary and alternative therapy associated with higher effectiveness and reduced toxicity [10]. Several medicinal plants, still unknown to the scientific fraternity, have already been reported in ethnomedicinal data [11]. Nowadays, a number of drugs that are commercially available have also been derived either from traditional medicines or from the plants used by ethnic groups. Recently, nanoscience and herbal science have been merged to improve the efficacy and reduce the toxicity of plant products as therapeutic agents.

10.2 DEVELOPMENT OF HERBAL MEDICINES

Ancient civilizations employed herbal medicines since time immemorial, however, the modern scientific exploration of plant products is the need of the hour for the development of these herbal medicines. A scientific approach to this problem has lately been introduced, and the data show that for the majority of phyto-products, the significant lack of knowledge needs to be addressed before their efficacy can be confirmed. However, although herbal medicines are exposed to thorough developmental procedures in some countries, this is not the case everywhere. For example, Germany markets herbal products as "phyto-medicines," which follow the same norms for quality, safety, and efficacy as other synthetic drugs. On the other hand, most herbal products in the US are marketed and standardized as nutritional supplements, falling into a product category for which no prior approval based on the above criteria is required.

According to the European Commission, medicinal products in all Member States, need consent before going to the market. These herbal medicinal products principally follow the general guidelines for medicines as placed by the national medicine laws. There is a particular definition available for herbal medicinal products in accordance with the EU Guideline under the name "Quality of Herbal Medicinal Products." This also covers plants, their parts, and even their preparations having therapeutic values. Several classes of medicinal products consisting of herbal preparations are present or under production. Hence, the majority of newly launched plant-derived herbal medicines are novel and symbolize tremendous involvement in the advancement of existing medicines.

Since medicinal plants constitute an exemplary source of biologically active phytochemicals, herbal medicines, with high therapeutic efficacy and devoid of adverse side effects, are considered as an informative tool in the field of drug development. Currently, reports on the bioactivities of several medicinal plants are rising enormously. Nevertheless, it is not feasible to indicate the activity of a single component from a mixture, since the activity of plant extracts is due to the synergistic effect of diverse range of phytochemicals present in it. It is well known that herbal medicines have a range of active constituents and all the components offer a synergistic effect and thus improve their medicinal value. However, these herbal drugs are less effective for instant medicinal applications as the majority of plant-derived medicines have less bioavailability and a greater rate of disposal from the system due to their low solubility, leading to the requirement for repetitive or increased dose administration. Due to these drawbacks, herbal medicines are less effective candidates for medicinal applications. Hence, the development of herbal products in order to add value to these products by improving their efficacy and bioavailability is one of the main areas of plant research these days.

Recently, nano-based drugs have emerged as a new trend in the development of herbal medicines, especially, the plant-derived production of nanoparticles is gaining immense significance [12]. Plant products with self-fabricated nanoparticles are emerging as the best examples of drug development, in which phytoconstituents affect the efficiency of the formation of nanoparticles and their properties and the nanoparticles enhance the bio-efficacy of the plant.

However, nanoparticles have several applications including disease diagnostics, but their use is limited in the development of herbal formulations. Conventional drugs are not as effective as the drug bound to nanoparticles. Nanoparticles show a higher effect by increasing the half-life and bioavailability, and convey a higher concentration to the site of action [13]. The surface characteristics and size of the drug nanoparticles can be modified to obtain the desired characteristics for desirable effects [14]. The nanoparticle bound with drug is free to circulate and thus can minimize the side effects of the drug enabling its greatest effect at the required site. As the drug bound to nanoparticle has an increased surface area the drug loading is high as compared to conventional drugs [15]. Redispersion of nanoparticle-bound drugs is easy as compared to conventional drugs so liquids can penetrate deeply into organs. With these advantages, the phytoconstituents can show high efficacy and bioavailability.

10.3   GREEN SYNTHESIS OF NANOPARTICLES: AGNPS AND FENPS

Metal nanoparticles should be synthesized through the development of reliable and environment-friendly techniques only. To fulfill this need, the living plants play an important role in the synthesis of nanoparticles (Figure 10.1). There are many advantages to the green synthesis of metal nanoparticles. The green synthesis technique is non-toxic, fast, takes place at ambient temperature, and is low cost. Plants and many plant parts are used for the synthesis of metal nanoparticles. More importantly, from the commercialization point of view, the plant system is an advantageous non-pathogenic biological system for green synthesis of metal nanoparticles.

The evolution of green approaches for the production of nanoparticles has been accomplished by a number of processes [16]. However, employing plants is a more desirable approach since it is fast, economical, eco-friendly, and a one-pot method for biological synthesis of nanoparticles, enabling efficient large-scale synthesis [17]. In very recent

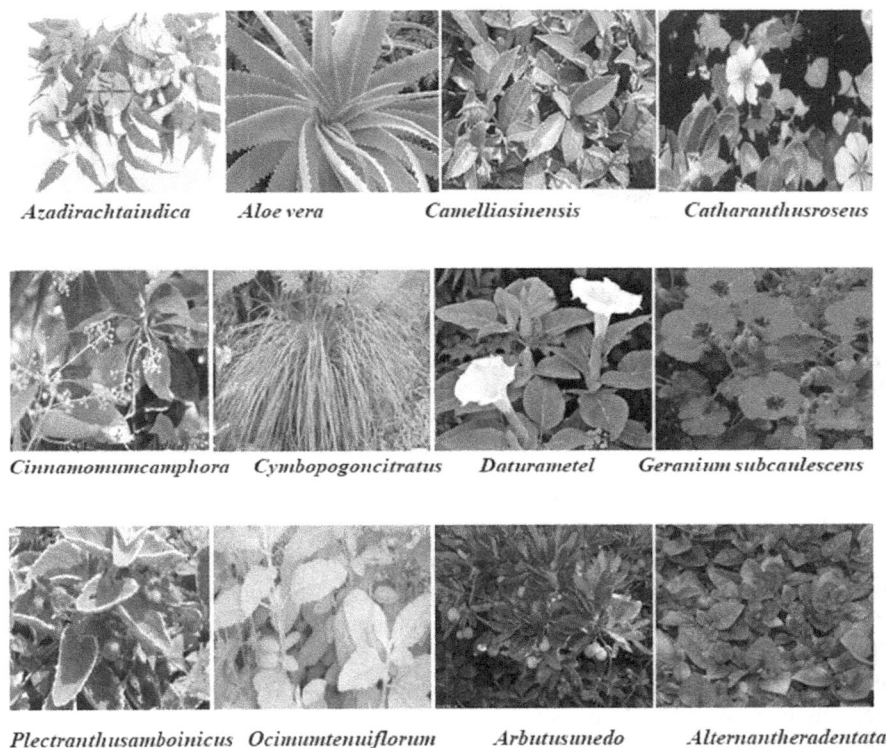

FIGURE 10.1   Different plants used for metal nanoparticle synthesis.

times, phyto-nano-synthesis has also gained immense recognition and is developing into a significant area of nanoscience involving green synthesis approaches [18].

In plant-mediated nanoparticle synthesis, biomolecules of plants play a dual role as both capping agents for stabilization as well as reducing agents. Therefore, the plant-derived nano-drugs hold a promising future with enhanced bioactivity, overcoming the usual drawbacks of low solubility and reduced bioavailability of herbal drugs. The possible mode of nanoparticle production as a bottom-up method is illustrated in Figure 10.2.

Usually, plant-derived biosynthesis includes the rapid reaction between aqueous plant extract and the corresponding metal salt in an aqueous medium at room temperature. The syntheses of several metal nanoparticles, such as those of Ag, Au, etc. have been carried out in this manner. Phytoconstituents as well as proteins are the chief biological molecules accountable for bringing about the reaction involving the dual function of capping and reducing agents. These secondary metabolites also function as stabilizing agents after the biosynthesis of plant-derived nanoparticles. This mode of biosynthesis is performed at favorable room temperature and in a simple way. The temperature, pH, type of metal salt, and concentration of phytoconstituents responsible for the reduction of metal ions have been found to influence the amount, rate, and properties of synthesized nanoparticles [19].

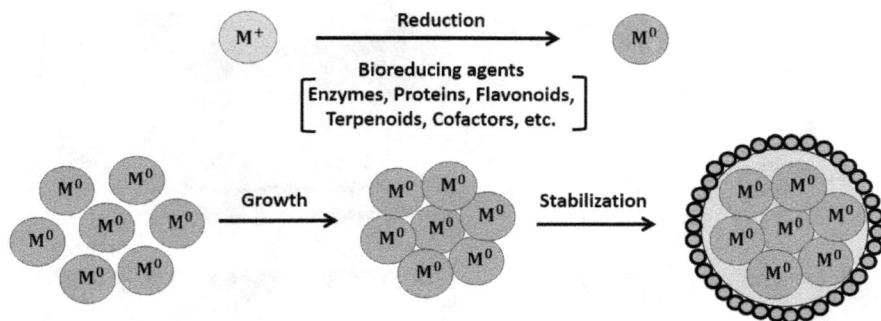

FIGURE 10.2   Plant-mediated nanoparticle synthesis.

AgNPs have become the most popular of all metal nanoparticles due to their several unique characteristics and enormous applicability in a wide range of areas [20]. Silver-derived antimicrobial agents are well known. The antimicrobial efficiency of silver is based on its cation Ag^+, which forms a strong bond with electron-donating groups found in biomolecules including oxygen, sulfur, or nitrogen. Thus, silver-based nanoparticles possessing increased surface area-to-volume ratio behave more efficiently and display enhanced attributes as compared to the larger metal molecules. The mechanistic approach of such biosynthesized nanoparticles is that they undergo biological reduction of Ag^+ to Ag^0, subsequently leading to self-assembly and colloidal aggregation [21]. As a result, these AgNPs command high applicability in the biomedical and medical fields.

The green approach of FeNP synthesis employing extract of green tea has evolved as a facile, affordable, and eco-friendly method recently. The majority of the data show green tea leaves being employed for FeNP synthesis, however inadequate information exists regarding the exploitation of other plants possessing substantial amounts of antioxidants, although lately, the antioxidant potential of some tree leaves has been evaluated to determine their feasibility [22].

10.4   CHARACTERIZATION OF PHYTOFABRICATED NANOPARTICLES

Nanoparticles have typically distinct characteristics of shape, size, disparity, surface area, and homogeneity. The common techniques of characterizing nanoparticles are: UV-visible spectroscopy, SEM (scanning electron microscopy), TEM (transmission electron microscopy), EPMA (electron probe microanalysis), FTIR (Fourier transform infrared spectroscopy), LIBS (laser-induced breakdown spectroscopy), and XRD (X-ray diffraction).

10.4.1   UV–VISIBLE SPECTROSCOPY

This is the most widely used technique for structural characterization of nanoparticles [23]. The plasmon bandwidth is found to follow the predicted behavior as it increases with decreasing size in the intrinsic size region (mean diameter less than 25 nm), and also increases with increasing size in the extrinsic size region (mean diameter larger than 25 nm). Hence, metal nanoparticles exhibit a particular surface plasmon absorbance bandwidth for a particular surface area of nanoparticles.

10.4.2 ELECTRON PROBE MICROANALYSIS (EPMA)

EPMA is a conventional technique used for the characterization of nano-sized substances. It helps in producing images of a surface by probing and scanning the sample physically. Elemental analysis of metal nanoparticles can also be performed if EDS (energy-dispersive spectroscopy) is coupled with EPMA.

10.4.3 X-RAY DIFFRACTION (XRD)

XRD is used for phase detection and analysis of nanoparticle crystal structure. X-rays enter the nanomaterial and the resultant diffraction pattern obtained is taken in comparison to standards in order to derive the structural data. Each crystalline solid has its individual distinctive XRPD pattern that may be employed as a "fingerprint" for its detection.

10.4.4 FOURIER TRANSFORM INFRARED SPECTROSCOPY (FTIR)

This is a remarkable tool for the identification of functional groups of phyto-constituents responsible for the bio-reduction of metal ions into metal nanoparticles. FTIR is used to detect the probable biomolecules bringing about a reduction of metal precursors to resultant nanoparticles and their metal capping ability. FTIR spectra provide data about chemical transformations in the bio-reduced functional groups.

10.4.5 LASER-INDUCED BREAKDOWN SPECTROSCOPY (LIBS)

LIBS is a form of atomic emission spectroscopy, which employs an extremely energetic laser pulse as the source of excitation. In principle, LIBS can identify any matter irrespective of its physical state, namely, solid, liquid, or gaseous. Since all elements generate light comprising distinct frequencies on reaching an excitation state at adequately high temperatures, LIBS can identify all elements, and is restricted simply by the laser power and the wavelength range and sensitivity of the detector and spectrograph.

10.5 IMPACT OF PLANT-MEDIATED NANOPARTICLES ON THE THERAPEUTIC EFFICACY OF MEDICINAL PLANTS

Plant extracts may act as both reducing agents and stabilizing/capping agents in the synthesis of nanoparticles. The origin of the plant is reported to influence the characteristics of nanoparticles [24]. This is because extracts of different plants have different combinations and concentrations of phyto-reducing agents. These plant-fabricated nanoparticles combined with phytoconstituents could be exploited for the development of value-added herbal medicinal products. Functional groups of phyto- constituents play an important role in the synthesis of nanoparticles, as a number of researchers have pointed toward different functional groups involved in the reduction of different metal ions to synthesize nanoparticles (Figure 10.3, Table 10.1).

Hence, phyto-synthesized nanoparticles are of great medicinal importance since they present the greatest suitability with bioactive phyto-molecules [35]. These studies will help budding nano-biotechnologists to harness biological systems for the synthesis of metal nanoparticles for different commercial applications in the medicine field.

Since nanoparticles act as a primary source of soluble metal, the combination of nanoparticles with natural products could be helpful in increasing the bioavailability of plant products. The synthesis of metal nanoparticles using plants is of increasing

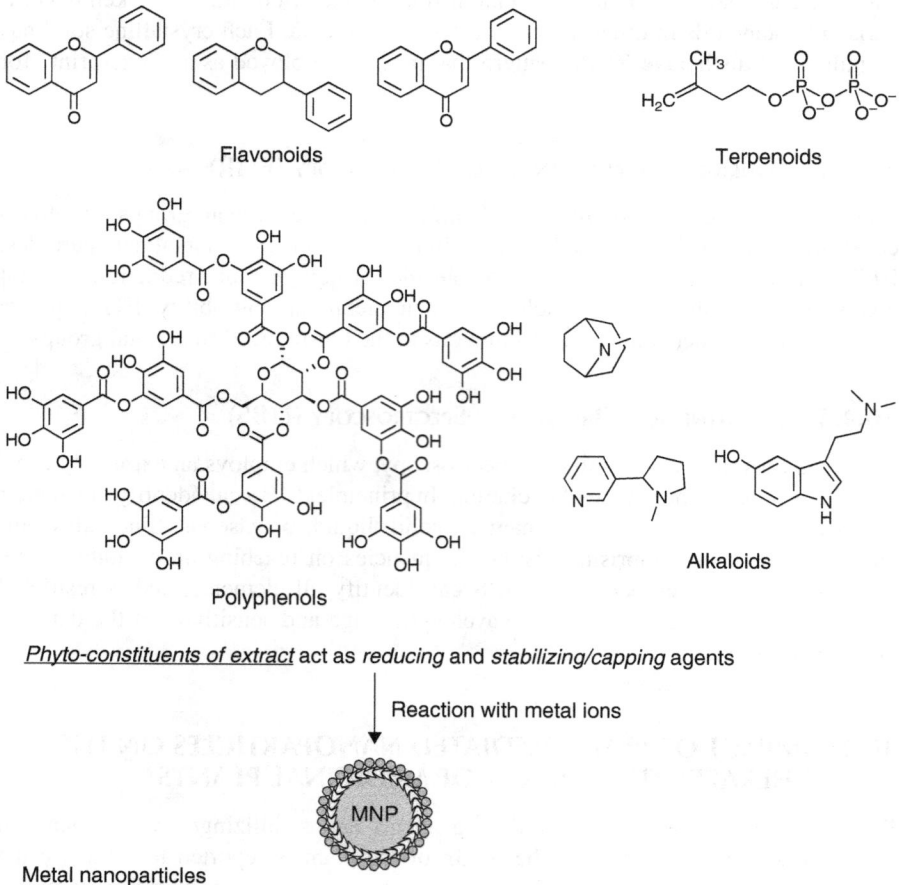

FIGURE 10.3 Phytoconstituents as bio-reducing agents of metal ions into metal nanoparticles.

TABLE 10.1
Some Examples of Plant Extracts Which Reduced Certain Metal Ions into Nanoparticles Due to the Presence of Certain Phytoconstituents

Plant and its part	Bioreduction of	Attributed to the presence of		References
C. camphora (Leaf extract)	—	Ag and Au ions	Terpenoids, polysaccharides	[25]
D. trifolium (Plant extract)	Flavones, phenolics	Ag ions	—	[26]
D. metel (Leaf extract)	—	Ag ions	Alkaloids, proteins and polysaccharides	[27]
O. sanctum (Leaf extract)	—	Ag ions	Ascorbic acid in great amount	[28]
C. fistula (Plant extract)	—	Au ions	Enhanced hypoglycemic activity than plant	[29]
A. indica (Leaf extract)	—	Ag and Au ions	Reducing sugars	[30]
P. graveolens (Leaf extract)	—	Ag and Au ions	Reducing sugars	[30]
T. vulgare (fruit extract)	—	Ag and Au ions	Compounds having carbonyl groups	[31]
Sorghum bran (aq. extract)	—	Ag and Fe ions	Reducing agents	[32]
T. chebula (aq. extract)	—	Fe ions	Natural antioxidants	[33]
Green tea (aq. extract)	—	Fe ions	Polyphenols	[34]

interest in the development of biomedicines as plants have great relevance in the bio-modification of metal nanoparticles and enhancing bioactivity (Figure 10.4).

For example, *Syzygium cumini* seed-derived AgNPs were found to exhibit enhanced antioxidant activity compared with the original plant extract due to desirable antioxidant compound absorption of the extract on the nanoparticle surface. Moreover, AgNPs synthesized from the leaf extract of *Ocimum sanctum* have been reported to possess immense antimicrobial potential against both Gram-positive (e.g., *Streptococcus aureus*) and Gram-negative (e.g., *E. coli*) bacteria. An aqueous extract of *Cassia fistula* was employed in the synthesis of AuNPs which demonstrated enhanced hypoglycemic activity compared with the whole plant as well as its bark. This enhanced efficacy was due to the extract-derived hypoglycemic agent aggregation on the surface of the metal nanoparticles. CuNPs synthesized from Magnolia leaf extract demonstrated increased antibacterial potential against the bacterial strain *E. coli* [36]. Plant-synthesized FeNPs exhibited better free radical scavenging ability and stability due to the presence of polyphenols or antioxidants in green tea extract which safeguard the particles from aggregating and oxidizing.

The present study is an interesting example of symbiosis of medicinal plants and nanoparticles, as the phytochemicals bound to nanoparticles have been claimed to be advantageous as compared with their original forms. Since nanoparticle-bound phytochemicals have an extended half-life in vivo, and longer circulation times therefore, they can convey a high concentration of that constituent to where it is needed. Nanoparticle-bound phytochemicals are easily suspended in liquids and hence easy to penetrate. Thus, medicinal plants help in the controlled synthesis of nanoparticles, and nanoparticles help them in increasing their bioactive potential.

However, the impact of nanoparticles on the bioactive potential of medicinal plants depends upon the efficiency of synthesis of nanoparticles and other factors which

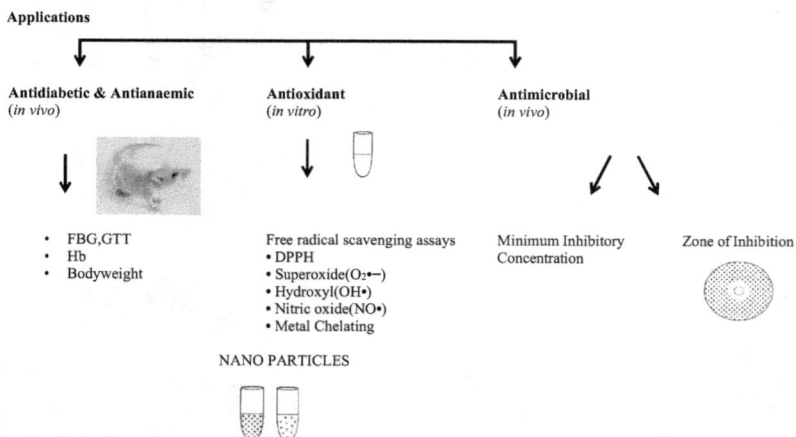

FIGURE 10.4   Therapeutic applications of plant-mediated nanoparticles.

play a key role in improving the efficiency of the synthesis and deciding the shape and size of nanoparticles, including pH, temperature, and the concentration of reducing phyto-agents. It is presumed that the increased bio-efficacy of medicinal plants in the presence of nanoparticles might be due to easy transportation of active constituents and their increased solubility and, hence, bioavailability.

Thus, combining nano-based herbal medicines with the conventional system of medicine is the need of the hour to develop better therapeutics with enhanced activity to combat long-term disorders such as diabetes, cancer, asthma, etc. Nanoparticles have gained significant importance in recent times due to their efficacy in applications for various herbal formulation areas, such as the development of herbal formulations containing antimicrobial agents, novel drug-delivery systems, lab and disease diagnostics, different cell labeling methods, use of isolated biomarkers, tagging of biologic materials, environmental factors, and in the pharmaceutical industry [37]. Research and development in this area is developing rapidly around the world. The primary outcome of this initiative is the rapid, precise, and secure synthesis of nanoparticles employing innovative processes that are preferably environmentally friendly.

Metal nanoparticles have attracted much attention because of their unique attributes for various applications, especially biomedical applications. In recent times, there has been an upsurge of interest in silver (Ag) as well as iron (Fe) nanoparticles due to their varied applications. With the increasing demand for AgNPs and FeNPs for different applications, including biomedical, efforts have been made in synthesizing these nanoparticles using easy, neat, safe, and quick methods. The oxidation–reduction and cellular respiration occurring in plants and animals uses iron at the active site of many redox enzymes. It also plays a role as a structural component as well as cofactor for different enzymes.

Since the majority of the cited methods include more than one step, require high energy, small conversion of matter, difficulty in refining, and use of harmful chemicals, cost-effective, commercially reproducible, and eco-friendly methods of synthesizing nanoparticles remain an unmet challenge.

Green synthesis offers an improvement over physical and chemical methods as it is cheap, environmentally benign, commercially viable, and requires no use of high energy, temperature, pressure, or harmful chemicals [38]. The green approach to synthesize nanoparticles using biological microorganisms, enzymes, or plant extracts has developed as an alternative to chemical synthesis.

Further studies are required to probe the role of different natural products during the synthesis of nanoparticles. Tree leaves, particularly fruit tree leaves, possess high levels of natural antioxidants due to the presence of polyphenols which enhance antioxidant attributes in the majority. Thus, phyto-antioxidants are not only reported for their known benefits on human well-being, but are also supposed to be better reducing agents for the synthesis of nanoparticles [39].

Herbal drugs have recently gained more attention because of their potential to treat almost all diseases. However, several problems such as poor solubility, poor bioavailability, low oral absorption, instability, and unpredictable toxicity of herbal medicines limit their use. It may therefore be stated that these nanoparticles are clean, easy, non-toxic, biodegradable, economically viable, as well as an eco-friendly green approach for synthesizing AgNPs and FeNPs using leaf extracts of medicinally important plants. In order to overcome such problems, nanoparticles can play a vital role. Hence, different nanoparticles, including polymeric nanoparticles, liposomes, proliposomes, solid lipid nanoparticles, and microemulsions, showcase potential utilization to deliver herbal medicines with better therapeutic results.

10.5.1 THERAPEUTIC APPLICATIONS OF NANOPARTICLES IN PATHOLOGICAL DISORDERS

Therapeutic applications of nanotechnology in medicine is known as nanomedicine. Nanomedicines have small particle size, large surface area to mass ratio, and thus affect the physicochemical properties and biological activities of drugs. These unique properties of nanomedicine can be utilized to overcome the limitations associated with the use of conventional therapeutic agents and diagnostic agents. In past decades, nanomedicine-based therapeutic pharmaceutical products have become increasingly available in commercial products. More than 100 pharmaceutical companies across the world have developed nanomedicines to be used in the treatment of various disease conditions. The most common nanotechnology-based products are liposomal drug-delivery systems and drug-conjugates with varying polymers, dendrimers, micelles, and inorganic nanoparticles [40]. Current therapeutic applications of nanotechnology in the treatment of various disease conditions are described below.

10.5.2 INFECTIOUS DISEASES

The current use of antimicrobial agents has numerous disadvantages such as poor concentration at the site of infection, risk of resistance, and associated toxicity. These could be overcome by the application of nanotechnology in the preparation of dosage forms [41,42]. One of the fluoroquinolone classes of antimicrobial agents, ciprofloxacin-inhaled liposomal formulation, is designed to provide sustained action for up to 24 hours and with lower associated side effects compared to conventional oral administration [43]. The systemic side effects of antifungal agent amphotericin B are also reduced by its liposomal drug-delivery system in the treatment of histoplasmosis associated with AIDS [44].

Viral structures such as viral glycoprotein toll-like receptors are prepared in the form of virosomal vaccines, and at present two such preparations have been approved in the treatment of hepatitis A and influenza [44]. Antimicrobial activity has been also established by in vitro methods using nanoparticles of fluconazole, ampicillin, and polymyxin B against *Candida*, *E.coli*, and *P. aeruginosa*, respectively [45,46] (Table 10.2).

10.5.3 CANCER

Caner is largest causes of mortality worldwide. Current treatment approaches include chemotherapy, radiotherapy, and surgery. It is well established that chemotherapy is associated with numerous systemic adverse drug reactions (ADRs), including gastric

TABLE 10.2
Therapeutics of Nanotechnology in the Treatment of Infectious Diseases

Nanotechnology	Company	Therapeutic indication
Liposomal amphotericin B	Enzon	i.v. for fungal infection
Liposomal vaccine	Berna Biotech	i.m. for hepatitis A and influenza
Polymeric (PEG) interferon 2a	Nektar Hoffmann-La Roche	s.c. for hepatitis B and C

ulcers and alopecia. Another challenge is the development of resistance due to efflux mechanisms. Ove the past few years these limitations have been overcome through the use of a nanoparticle-based cancer targeting delivery system of drugs. This system targets the tumor only, rather than normal cells, thus minimizing the ADRs [47,48]. Tumors have leaky vasculatures and networks of lymphatic vessels making them suitable to retain the nanoparticles of drugs [49]. Table 10.3 provides a list of therapeutic agents incorporating nanotechnology that have undergone clinical trials or been approved by the international FDAs of various countries for the treatment of cancer.

10.5.4 AUTOIMMUNE DISEASES

Some autoimmune disease treatments include the use of nanotechnology-mediated drug-delivery systems. These include rheumatoid arthritis (RA), multiple sclerosis (MS), AIDS, and amyloidosis. This can reduce the ADRs associated with the long-term treatment of such diseases. The nano-formulation of certolizumab (a TNF inhibitor) with PEG raises the $t_{1/2}$ to 2 weeks and has shown good results in the long-term management of RA. It alleviated the process of bone resorption and destruction [50]. Table 10.4 provides a list of therapeutic agents incorporating nanotechnology that have undergone clinical trials or been approved by the international FDAs of various countries for the treatment of autoimmune diseases.

10.5.5 NEURODEGENERATIVE DISEASES

These diseases are characterized by the progressive loss of neurons and their functions leading to neuronal death. With a gradual loss of neurons, patients with various neurodegenerative diseases such as Alzheimer's disease (AD), Parkinson's disease (PD), and multiple sclerosis (PD) have numerous symptoms related to movement and dementia. To treat such diseases, drugs must cross the blood–brain barrier (BBB) to enter the brain. The BBB is highly selective and only a small portion of therapeutic drugs can enter the brain. Thus, this results in the use of a large dose of drug and hence increases the chances of ADRs. In such diseases, nanoparticle-based therapeutic approaches focus on the delivery of drug at the sites affected in the brain after crossing the BBB [51,52]. Some drugs have been reformulated with the use of

TABLE 10.3
Therapeutics of Nanotechnology in the Treatment of Cancer

Nanotechnology	Company	Therapeutic indication
Liposomal cytarabine	Depocyt	Malignant meningitis
Liposomal daunorubicin	Gilead Sciences	i.v. HIV-related Kaposi's sarcoma
Liposomal doxorubicin	Zeneus	i.v. metastatic breast cancer
		i.m. for ovarian and breast cancer
Liposomal vincristine sulfate	Spectrum	i.v. acute lymphoblastic leukemia (ALL)
Liposomal mifamurtide	Takeda	i.v. non-metastasizing osteosarcoma
Liposomal Irinotecan	Ipsen	i.v. pancreatic and colorectal cancer
Polymeric methoxy PEG poly taxol	Samyang	i.v. metastatic breast cancer
Polymeric PEG L-asparaginase	Enzon	i.v. and i.m. acute lymphoblastic leukemia (ALL)
Protein-bound paclitaxel	AstraZeneca Celgene	i.v. metastatic breast cancer i.v. non-small lung cancer i.v. pancreatic cancer
Protein conjugate of trastuzumab	Genentech	i.v. HER2+ metastatic breast cancer
Metallic nanoparticles of iron oxide coated with amino silane	MagForceAG	Magnetic hyperthermia therapy for glioblastoma, prostate cancer, and pancreatic cancer
Polymeric doxorubicin	Bio-Alliance Pharma	i.a. hepatocellular carcinoma

TABLE 10.4
Therapeutics of Nanotechnology in the Treatment of Autoimmune Disorders

Nanotechnology	Company	Therapeutic indication
Lipid based nonliposome of transthyretin-targeted siRNA	Alnylam	i.v. transthyretin-mediated amyloidosis
Polymer-based glatiramer	Novartis	s.c. in multiple sclerosis
Protein-bound conjugate of interferon β-1a	Biogen	s.c. in multiple sclerosis
Polymeric L-glutamic acid, L-alanine L-lysine and L-tyrosine copolymer	Teva	s.c. in multiple sclerosis
Polymeric anti-TNF-α antibody	Nektar	Rheumatoid arthritis and Crohn's disease

nanotechnology and evaluated in preclinical animal models including cholinesterase inhibitors such as donepezil, tacrine, and rivastigmine, as well as the N-methyl-d-aspartate (NMDA) receptor antagonist memantine [53–56]. These agents have shown improved penetration through the BBB as compared to free drug administration. In the same way, in the treatment of PD, nanoparticles of ropirinol, bromocriptine, and apomorphine have shown improved release profile and reduced side effects when compared to conventional drugs [57,58].

10.5.6 CARDIOVASCULAR (CV) AND BLOOD DISEASE

CV diseases affect the vascular system, kidneys, and various peripheral arteries. Well-known diseases of this type include arteriosclerosis, atherosclerosis, and peripheral vascular disease. These are more common in patients with diabetes mellitus as micro-vascular complications. CV and blood diseases are leading causes of death world-wide [58].

Liposomes of the immunosuppressant drug sirolimus formed by lecithin and cholesterol and a coating of chitosan have been developed for the prevention of vas-cular restenosis [59]. Increased bioavailability is obtained also with the nanoparticle formulation of carvedilol, an agent used in the treatment of myocardial infarction, congestive cardiac failure, and left ventricular dysfunction [60]. Table 10.5 provides a list of therapeutic agents incorporating nanotechnology that have undergone clin-ical trials or been approved by the international FDAs of various countries for the treatment of cardiovascular and blood disorders.

10.5.7 OPHTHALMIC DISEASES

At present, nanotechnological interventions with mydriatics, miotics, cycloplegics, antiinflammatory agents, antiinfectives, and diagnostic agents have been evaluated for their ophthalmic therapeutic role. However, there are certain limitations for the penetration of drugs through various layers of the eye [61]. Nanoparticles of various drugs showed increased bioavailability by providing better penetration through eye membranes. Liposomal and polymeric nanoparticle formulations of antiinflamma-tory agents have been evaluated in ophthalmic eye models [62]. Nanoparticles of brimonidine were used in the treatment of glaucoma [63–65].

10.5.8 RESPIRATORY DISEASES

These diseases include asthma, chronic obstructive pulmonary disease (COPD), bronchitis, cystic fibrosis, pulmonary tuberculosis, and bronchiectasis [66]. There is

TABLE 10.5
Therapeutics of Nanotechnology in the Treatment of CV and Blood Disorders

Nanotechnology	Company	Therapeutic indication
Nanocrystalline fenofibrate	Abbott, Elan	Orally for hyperlipidemia
Nanocrystalline sirolimus	Elan, Wyeth	Orally as an immunosuppressant
Protein-bound conjugate of factor VIII	Baxalta	i.v. for hemophilia A
Protein bound conjugate of interferon IX	Novo Nordisk	i.v. for hemophilia B

no effective cure for such diseases. The use of aerosols containing sympathomimetics and steroids has been well established in their treatment. Polyamidoamine dendrimers made with beclomethasone dipropionate are used to treat asthma. They provide better antiinflammatory action compared to conventional aerosol preparations. Polymer-based nanoparticles and carbon-based nanoparticles have been used also for vaccine delivery in the respiratory system [67].

10.5.9 TISSUE REGENERATION THERAPY

Nanoparticles of collagen, gelatin, chitosan, and albumin have been used for bone regeneration [68]. Bone formation is enhanced by the stimulation of osteoblast activity through the delivery of nanoparticles containing growth factors [69]. Polymeric nanoparticles of bisphosphonates also have been used in the treatment of osteoporosis as they stimulate osteoclast apoptosis [70]. Nanoparticles of hydroxyapatite, demineralized bone matrix, and carboxymethyl cellulose-containing products were approved by the FDA in 2009 for use as an osteoinductive bone graft substitute [71].

10.6 OTHER INDICATIONS

Intramuscular nanocrystalline paliperidone palmitate (Elan, Johnson & Johnson) is an antipsychotic agents under phase III clinical trial. It is a long-acting drug with reduced frequency of drug administration [71]. Propofol, an i.v. general anesthetic, is under phase III clinical trial (Skype Pharma) to be used with various surgeries. Its composition consists of insoluble drug-delivery microdroplets (IDD-D) of propofol. Phoenix conduced phase III trials for i.v. PEG-uricase for the treatment of severe hyperuricemia from gout with reduced side effects. Metallic nanoparticles of iron to be given via the parenteral route have been approved for the treatment of iron-deficiency anemia and anemia associated with chronic kidney injury by the USFDA in 2009 and the European FDA in 2012 [72].

10.6.1 NANOPARTICLES IN CANCER DIAGNOSIS AND THERAPEUTICS

Nanoparticles are a fascinating tool used in biomedical applications due to their active surface and drug encapsulation potential. They have been enormously researched for cancer diagnoses and therapeutics. The higher mortality rate in various types of cancers is mainly due to delayed diagnosis. Nanoparticles could contribute to sensitive and specific diagnostic methods by detecting cancer markers. Currently, a tissue biopsy is a clinically approved standard technique to diagnose cancer. Cancer heterogeneity is a challenge in tissue biopsy because tumor sample extraction may not provide detailed information about the patient's primary and secondary tumors. In clinical practice, the cancer biomarkers are also analyzed using body fluids. The collection of body fluids is easy as they are available in various forms, such as blood, urine, and cerebrospinal fluid. Despite the availability of body fluids and their easy detection, this method of diagnosis lacks the sensitivity of cancer markers such as carcinoembryonic and prostate-specific antigens. Cancer marker analysis reveals the presence of cancer in the body but does not provide any prognosis information. Nanotechnology could play an important role in detecting cancer biomarkers and also provide detailed information about cancer prognosis.

The smaller gold nanoparticles have shown lower systemic toxicity with faster clearance, and this nanosystem showed significant accumulation in the tumor site. These accumulated gold nanoparticles could be visualized using various advanced imaging tools. Melanie et al. functionalized gold nanoparticles with heat shock protein 70 antibody and visualized the accumulated nanoparticles in the tumor using spectral CT analysis [73]. Zhang et al. demonstrated that gold nanoparticles with a zwitterionic surface could form large aggregates in the tumor's acidic environment. This aggregated form of gold could be imaged using photoacoustic imaging [74]. Near-infrared-absorbing organic photosensitizers have excellent optical properties but poor photostability and very low bioavailability. The encapsulation of this dye into nanosystems enhances its photostability and bioavailability. The near-infrared fluorescence emission could be imaged with in vivo imaging systems [75].

Estrogen, progesterone, and epidermal growth factor receptors are the cancer cell membrane biomarkers used in breast cancer clinical diagnosis. The nanoparticle-linked immunosorbent assay could provide a selective and sensitive diagnosis method [76]. The functionalized nanoparticles could be optimized for enhanced binding potential to cancer biomarkers. The binding of fluorescent nanoparticles to cancer markers alters the fluorescence intensity. The fluorescent-based detection methods are reliable and highly sensitive [77]. The immune-optomagnetic quantum dot-based assays could provide easy and accurate cancer detection without the need for sophisticated instruments [78]. The development of a simple nanoparticle-based immune assay could detect cancer markers using a nanoparticle–antibody conjugate in the assay [79].

Nanoparticle-mediated therapy is an extensively researched cancer treatment modality that includes efficient drug delivery at the tumor site and photoablation of tumor by inducing hyperthermia or generating free radicals. Nanoparticles are smart carriers for hydrophobic and hydrophilic drugs due to their high drug-loading capacity of various nanosystem types. Matrix metalloproteinase (MMP)-mediated surface modification of liposomes could enhance the tumor-targeting efficacy of nanosystems in response to the tumor microenvironment [80]. Polymeric nanoparticles could form a stable colloidal with higher drug encapsulations, and this nanosystem could induce the anionic and cationic surface layer for advanced applications [81]. Kunho et al. developed PLGA nanoparticles with RGD ligand that help in the tumor-specific targeting of glioblastoma multiforme [82]. Surface modification could be achieved using a mucoadhesive polymer coating to enable the maximum residential time toward cancer cells and the tumor microenvironment [83]. The therapeutic drugs could tag to the metallic nanoparticles' surfaces for targeted drug delivery to the tumor. Arokia et al. developed copper oxide nanoparticles encapsulating starch and functionalized with folic acid for targeting breast cancer cell lines. The folic acid and starch decorating copper oxide nanoparticles enhance nanoparticle penetration in cancer cells [84]. Vladimir et al. developed a metal-organic functionalized framework to deliver cytotoxic drugs [85].

Plasmonic and superparamagnetic nanoparticles are of interest in research as they produce heat and cause hyperthermia in tumors [86]. Gold nanoparticles are mainly focused on due to their plasmonic resonance phenomenon and are used in photothermal therapy. Branched gold nanoparticles could absorb NIR light and produce heat to kill cancer cells [87]. Sajid et al. developed anisotropic gold nanoparticles absorbing NIR light that are used for photothermal therapy and tumor imaging by CT scan [88]. Iron oxide nanoparticles are being extensively researched for magnetic hyperthermia to kill cancer cells [89]. Besides metallic nanosystems, the NIR dyes have been researched for photothermal and photodynamic therapies. ICG is an FDA-approved cyanine dye used in the clinic as a contrast agent in imaging [90]. Photostability and bioavailability are major challenges of NIR dye in biomedical applications. The entrapment of this dye could enhance the bioavailability and photostability of dyes. The dye could entrap plasmonic nanoparticles to enhance the nanosystems' and NIR dye's potential [91]. The nanoparticles also are combined with various treatment modalities such as chemotherapy and photothermal therapy by entrapping more than two therapeutic agents [92]. Developing new therapeutic formulations using nanoparticles could enhance the potential for improved therapeutic output.

REFERENCES

1. Bhattacharya D, Gupta RK. Nanotechnology and potential of microorganisms. *Critical Reviews in Biotechnology.* 2005 Oct–Dec;25(4):199–204. doi: 10.1080/07388550500361994
2. Gittins, D., Bethell, D., Schiffrin, D. et al. A nanometre-scale electronic switch consisting of a metal cluster and redox-addressable groups. *Nature* 408, 67–69 (2000). https://doi.org/10.1038/35040518
3. Darroudi,M., Ahmad, M. B., Abdullah. A. H., Ibrahim, N. A. (2011) Green synthesis and characterization of gelatin-based and sugar-reduced silver nanoparticles, *International Journal of Nanomedicine,* 6:, 569–574, DOI: 10.2147/IJN.S16867
4. Wise, K., Brasuel, M. (2011) The current state of engineered nanomaterials in consumer goods and waste streams: the need to develop nanoproperty-quantifiable sensors for monitoring engineered nanomaterials, *Nanotechnology, Science and Applications,* 4:, 73–86, DOI: 10.2147/NSA.S9039
5. Singh, T. P., Singh. O. M.. (2011). Phytochemical and pharmacological profile of Zanthoxylum armatum DC.-an overview.
6. Marcato, P. D., Durán, N. (2008) New aspects of nanopharmaceutical delivery systems. *Journal of Nanoscience and Nanotechnology* 8, no. 5: 2216–2229.
7. Yadav, D., Suri, S., Choudhary, A.A., Sikender, M., Hemant, K., Beg, N.M., et al. (2011) Novel approach: Herbal remedies and natural products in pharmaceutical science as nano drug delivery systems. *International Journal of Pharmacy and Technology* 3: 3092–3116.
8. Newman, D.J., Cragg, G.M. (2007) Natural products as sources of new drugs over the last 25 years. *Journal of Natural Products* 70: 461–477.
9. Butler, M.S. (2008) Natural products to drugs: Natural product derived compounds in clinical trials. *Natural Product Reports* 25: 475–516.

10. Ganesan, A. (2008) The impact of natural products upon modern drug discovery. *Current Opinion in Chemical Biology* 12: 306–317.
11. Shekhawat, M.S., Manokari, M., Kannan, N., Revathi, J. (2013) Synthesis of silver nanoparticles for *Cardiospermum helicobacum* leaf extract. *Phytopharmacology Journal* 2: 15–20.
12. Charlish, P. (2008) Traditional remedies: Latter day medicines. *Scrip World Pharmaceutical News* 3351: 31–34.
13. Singh, R.P., Singh, S.G., Naik, H., Jain, D., Bisla, S. (2011) Herbal excipients in novel drug delivery system. *International Journal of Comprehensive Pharmacology* 2: 1–7.
14. Bairwa, N.K., Sethiya, N.K., Mishra, S.H. (2010) Protective effect of stem bark of Ceibapentandra linn. against paracetamol-induced hepatotoxicity in rats. *Pharmacognosy Research* 2: 26–30.
15. Wu, X.Y., Lee, P.I. (1993) Preparation and characterization of thermal- and pH-sensitive nanospheres. *Pharmaceutical Research* 10: 1544–1547.
16. Elzoghby, A., Samy, W., Elgindy, N. (2012) Protein-based nanocarriers as promising drug and gene delivery systems. *Journal of Controlled Release* 161(1): 38–49.
17. Fang, R., Hao, R., Wu, X., Li, Q., Leng, X., Jing, H. (2011) Bovine serum albumin nanoparticles promotes the stability of quercetin in simulated intestinal fluid. *Journal of Agricultural and Food Chemistry* 59: 6292–6298.
18. Fessi, H., Puisieux, F., Devissaguet, J.P., Ammoury, N., Benita, S. (1989) Nano-capsule formation by interfacial polymer deposition following solvent displacement. *International Journal of Pharmaceutics* 55: 1–4.
19. Formica, J.V., Regelson, W. (1995) Review of the biology of quercetin and related bioflavonoids. *Food and Chemical Toxicology* 33: 1061–1080.
20. Fukuda, K., Hibiya, Y., Mutoh, M., Koshiji, M., Akao, S., Fujiwara, H. (1999) Inhibition of activation protein 1 activity by berberine in human hepatoma cells. *Planta Medica* 65: 381–383.
21. Gleave, M., Bruchovsky, N., Goldenberg, S.L., Rennie, P. (1998) Intermittent androgen suppression for prostate cancer: rationale and clinical experience. *European Urology* 34: 37–41.
22. Goyal, A., Kumar, S., Nagpal, M., Singh, I., Arora, S. (2011) Potential of novel drug delivery system for herbal system for herbal drugs. *Indian Journal of Pharmaceutical Education and Research* 45: 225–235.
23. Gupta, V.K., Karar, P.K., Ramesh, S., Misra, S.P., Gupta, A. (2010) Nanoparticle formulation for hydrophilic and hydrophobic drugs. *International Journal of Research in Pharmaceutical Sciences* 1: 163–169.
24. Li, H.-Y., Wang, D.-P., Zhang, T.-M., Ren, H.-L., Xu, F.-Y., Zhao, Z.-G. (2010) Improving the anti-tumor effect of genistein with a biocompatible superparamagnetic drug delivery system. *Journal of Nanoscience and Nanotechnology* 10(4): 2325–2331.
25. Huang, J., Li, Q., Sun, D., Lu, Y., Su, Y., Yang, X., ... & Chen, C. (2007). Biosynthesis of silver and gold nanoparticles by novel sundried Cinnamomum camphora leaf. *Nanotechnology*, 18(10), 105104.
26. Ahmad, P., Jaleel, C. A., Salem, M. A., Nabi, G., & Sharma, S. (2010) Roles of enzymatic and nonenzymatic antioxidants in plants during abiotic stress, *Critical Reviews in Biotechnology*, 30(3), 161–175, DOI: 10.3109/07388550903524243
27. Kesharwani, J., Yoon, K. Y., Hwang, J., & Rai, M. (2009). Phytofabrication of silver nanoparticles by leaf extract of Datura metel: hypothetical mechanism involved in synthesis. *Journal of Bionanoscience*, 3(1), 39–44.
28. Mallikarjuna, K., Narasimha, G., Dillip, G. R., Praveen, B., Shreedhar, B., Lakshmi, C. S., ... & Raju, B. D. P. (2011). Green synthesis of silver nanoparticles using

Ocimum leaf extract and their characterization. Digest *Journal of Nanomaterials and Biostructures*, 6(1), 181–186.

29. Daisy, P., & Saipriya, K. (2012). Biochemical analysis of Cassia fistula aqueous extract and phytochemically synthesized gold nanoparticles as hypoglycemic treatment for diabetes mellitus. *International Journal of Nanomedicine*, 1189–1202.

30. Shankar, S. S., Rai, A., Ahmad, A., & Sastry, M. (2004). Rapid synthesis of Au, Ag, and bimetallic Au core–Ag shell nanoparticles using Neem (Azadirachta indica) leaf broth. *Journal of Colloid and Interface Science*, 275(2), 496–502.

31. Dubey, S. P., Lahtinen, M., & Sillanpää, M. (2010). Tansy fruit mediated greener synthesis of silver and gold nanoparticles. *Process Biochemistry*, 45(7), 1065–1071.

32. Njagi, E. C., Huang, H., Stafford, L., Genuino, H., Galindo, H. M., Collins, J. B., ... & Suib, S. L. (2011). Biosynthesis of iron and silver nanoparticles at room temperature using aqueous sorghum bran extracts. *Langmuir*, 27(1), 264–271.

33. Kumar, S., & Pandey, A. K. (2013). Chemistry and biological activities of flavonoids: an overview. *The scientific world journal*, 2013.

34. Shahwan, T., Sirriah, S. A., Nairat, M., Boyacı, E., Eroğlu, A. E., Scott, T. B., & Hallam, K. R. (2011). Green synthesis of iron nanoparticles and their application as a Fenton-like catalyst for the degradation of aqueous cationic and anionic dyes. *Chemical Engineering Journal*, 172(1), 258–266.

35. Kang, D.G., Oh, H., Sohn, E.J., Hur, T.Y., Lee, K.C., Kim, K.J., Kim, T.Y., Lee, H.S. (2004) Lithospermic acid B isolated from *Salvia miltiorrhiza* ameliorates ischemia/reperfusion-induced renal injury in rats. *Life Sciences* 75: 1801–1816.

36. Kiuchi, F., Goto, Y., Sugimoto, N., Akao, N., Kondo, K., Tsuda, Y. (1993) Nematocidal activity of turmeric: synergistic action of curcuminoids. *Chemical and Pharmaceutical Bulletin (Tokyo)* 41: 1640.

37. Kumar, A., Yadev, S.K., Pakade, Y.B., Singh, B., Yadev, S.C. (2010) Development of biodegradable nanoparticles for delivery of quercetin. *Colloids and Surface B, Biointerfaces* 15: 184–192.

38. Kumar, K., Rai, A.K. (2012) Miraculous therapeutic effects of herbal drugs using novel drug delivery systems. *International Research Journal of Pharmacy* 3: 2733.

39. Vollath, D. (2008) Nanomaterials—An introduction to synthesis, properties and application. *Environmental Engineering and Management Journal* 7: 865–870.

40. Hillaireau, H., Couvreur, P. (2009) Nanocarriers' entry into the cell: relevance to drug delivery. *Cellular and Molecular Life Sciences* 66(17): 2873–2896.

41. Sendi, P., Proctor, R.A. (2009) *Staphylococcus aureus* as an intracellular pathogen: the role of small colony variants. *Trends in Microbiology* 17(2): 54–58.

42. Zazo, H., Colino, C.I., Lanao, J.M. (2016) Current applications of nanoparticles in infectious diseases. *Journal of Controlled Release* 28(224): 86–102.

43. Cornely, O.A., Maertens, J., Bresnik, M., Ebrahimi, R., Ullmann, A.J., Bouza, E., Heussel, C.P., Lortholary, O., Rieger, C., Boehme, A., Aoun, M. (2007) Liposomal amphotericin b as initial therapy for invasive mold infection: a randomized trial comparing a high–loading dose regimen with standard dosing (AmBiLoad Trial). *Clinical Infectious Diseases* 44(10): 1289–1297.

44. Alipour, M., Halwani, M., Omri, A., Suntres, Z.E. (2008) Antimicrobial effectiveness of liposomal polymyxin B against resistant Gram-negative bacterial strains. *International Journal of Pharmaceutics* 355(1–2): 293–298.

45. Mishra, B.B., Patel, B.B., Tiwari, S. (2010) Colloidal nanocarriers: a review on formulation technology, types and applications toward targeted drug delivery. *Nanomedicine: Nanotechnology, Biology and Medicine* 6(1): 9–24.

46. Danhier, F., Feron, O., Préat, V. (2010) To exploit the tumor microenvironment: passive and active tumor targeting of nanocarriers for anti-cancer drug delivery. *Journal of Controlled Release* 148(2): 135–146.

47. Maeda, H., Bharate, G.Y., Daruwalla, J. (2009) Polymeric drugs for efficient tumor-targeted drug delivery based on EPR-effect. *European Journal of Pharmaceutics and Biopharmaceutics* 71(3): 409–419.

48. Yudoh, K., Karasawa, R., Masuko, K., Kato, T. (2009) Water-soluble fullerene (C60) inhibits the development of arthritis in the rat model of arthritis. *International Journal of Nanomedicine* 4: 217.

49. Wohlfart, S., Gelperina, S., Kreuter, J. (2012) Transport of drugs across the blood–brain barrier by nanoparticles. *Journal of Controlled Release* 161(2): 264–273.

50. Liu, H.L., Fan, C.H., Ting, C.Y., Yeh, C.K. (2014) Combining microbubbles and ultrasound for drug delivery to brain tumors: current progress and overview. *Theranostics* 4(4): 432.

51. Wilson, B., Samanta, M.K., Santhi, K., Kumar, K.P., Paramakrishnan, N., Suresh, B. (2008) Poly (n-butylcyanoacrylate) nanoparticles coated with polysorbate 80 for the targeted delivery of rivastigmine into the brain to treat Alzheimer's disease. *Brain Research* 20(1200):159–168.

52. Patel, H.N., Patel, P.M. (2013) Dendrimer applications—A review. *International Journal of Pharmaceutical and Biological Sciences* 4(2): 454–463.

53. Laserra, S., Basit, A., Sozio, P., Marinelli, L., Fornasari, E., Cacciatore, I., Ciulla, M., Türkez, H., Geyikoglu, F., Di Stefano, A. (2015) Solid lipid nanoparticles loaded with lipoyl–memantine codrug: Preparation and characterization. *International Journal of Pharmaceutics* 485(1–2): 183–191.

54. Saraiva, C., Praça, C., Ferreira, R., Santos, T., Ferreira, L., Bernardino, L. (2016) Nanoparticle-mediated brain drug delivery: overcoming blood–brain barrier to treat neurodegenerative diseases. *Journal of Controlled Release* 10(235): 34–47.

55. Hu, H., Qiao, Y., Meng, F., Liu, X., Ding, C. (2013) Enhanced apatite-forming ability and cytocompatibility of porous and nanostructured $TiO_2/CaSiO_3$ coating on titanium. *Colloids and Surfaces B: Biointerfaces* 101: 83–90.

56. Joshi, S.A., Chavhan, S.S., Sawant, K.K. (2010) Rivastigmine-loaded PLGA and PBCA nanoparticles: preparation, optimization, characterization, in vitro and pharmacodynamic studies. *European Journal of Pharmaceutics and Biopharmaceutics* 76(2): 189–199.

57. Neeland, I.J., Poirier, P., Despres, J.P. (2018) Cardiovascular and metabolic heterogeneity of obesity: Clinical challenges and implications for management. *Circulation* 137: 1391–1406

58. Haeri, A., Sadeghian, S., Rabbani, S., Anvari, M.S., Ghassemi, S., Radfar, F., Dadashzadeh, S. (2017) Effective attenuation of vascular restenosis following local delivery of chitosan decorated sirolimus liposomes. *Carbohydrate Polymers* 10(157):1461–1469.

59. Arzani, G., Haeri, A., Daeihamed, M., Bakhtiari-Kaboutaraki, H., Dadashzadeh, S. (2015) Niosomal carriers enhance oral bioavailability of carvedilol: effects of bile salt-enriched vesicles and carrier surface charge. *International Journal of Nanomedicine* 10: 4797.

60. Sharaf, M.G., Cetinel, S., Heckler, L., Damji, K., Unsworth, L., Montemagno, C. (2014) Nanotechnology-based approaches for ophthalmology applications: therapeutic and diagnostic strategies. *The Asia-Pacific Journal of Ophthalmology* 3(3): 172–180.

61. Abrego, G., Alvarado, H., Souto, E.B., Guevara, B., Bellowa, L.H., Parra, A., Calpena, A., Garcia, M.L. (2015) Biopharmaceutical profile of pranoprofen-loaded PLGA

nanoparticles containing hydrogels for ocular administration. *European Journal of Pharmaceutics and Biopharmaceutics* 95: 261–270.

62. El-Salamounia, N.S., Farida, R.M. (2016) Recent drug delivery systems for treatment of glaucoma. *Glaucoma* 1–13: 231.
63. Ibrahim, M.M., Abd-Elgawad, A.H., Soliman, O.A., Jablonski, M.M. (2015) Natural bioadhesive biodegradable nanoparticle-based topical ophthalmic formulations for management of glaucoma. *Translational Vision Science and Technology* 4: 12.
64. Cetinel, S., Montemagno, C. (2016) Nanotechnology applications for glaucoma. *The Asia-Pacific Journal of Ophthalmology* 5: 70–78.
65. Sugawara, E., Nikaido, H. (2014) Properties of AdeABC and AdeIJK efflux systems of *Acinetobacter baumannii* compared with those of the AcrAB-TolC system of *Escherichia coli*. *Antimicrobial Agents and Chemotherapy* 58: 7250–7257.
66. Mohamud, R., Xiang, S.D., Selomulya, C., Rolland, J.M., O'Hehir, R.E., Hardy, C.L., Plebanski, M. (2014) The effects of engineered nanoparticles on pulmonary immune homeostasis. *Drug Metabolism Reviews* 46: 176–190.
67. Alves Cardoso, D., Jansen, J.A., Leeuwenburgh, S.C. (2012) Synthesis and application of nanostructured calcium phosphate ceramics for bone regeneration. *Journal of Biomedical Materials Research Part B: Applied Biomaterials* 100(8): 2316–2326.
68. Park, J.S., Yi, S.W., Kim, H.J., Kim, S.M., Park, K.H. (2016) Regulation of cell signaling factors using PLGA nanoparticles coated/loaded with genes and proteins for osteogenesis of human mesenchymal stem cells. *ACS Applied Materials and Interfaces* 8(44): 30387–30397.
69. Loh, X.J., Lee, T.C., Dou, Q., Deen, G.R. (2016) Utilizing inorganic nanocarriers for gene delivery. *Biomaterials Science* 4(1): 70–86.
70. Kurien, T., Pearson, R.G., Scammell, B.E. (2013) Bone graft substitutes currently available in orthopedic practice: the evidence for their use. *The Bone and Joint Journal* 95(5): 583–597.
71. Kaduk, J.A., Dmitrienko, A.O., Gindhart, A.M., Blanton, T.N. (2017) Crystal structure of paliperidonepalmitate (INVEGA SUSTENNA®), *C39H57FN4O4*. *Powder Diffraction* 32(4): 222.
72. Zhang, L., Gu, F.X., Chan, J.M., Wang, A.Z., Langer, R.S., Farokhzad, O.C. (2008) Nanoparticles in medicine: therapeutic applications and developments. *Clinical Pharmacology & Therapeutics* 83(5): 761–769.
73. Kimm, M.A., Shevtsov, M., Werner, C., Sievert, W., Zhiyuan, W., Schoppe, O., Menze, B.H., Rummeny, E.J., Proksa, R., Bystrova, O., Martynova, M., Multhoff, G., Stangl, S. (2020) Gold nanoparticle mediated multi-modal CT imaging of Hsp70 membrane-positive tumors. *Cancers (Basel)* 12: 1–17.
74. Zhang, R., Wang, L., Wang, X., Jia, Q., Chen, Z., Yang, Z., Ji, R., Tian, J., Wang, Z. (2020) Acid-induced in vivo assembly of gold nanoparticles for enhanced photoacoustic imaging-guided photothermal therapy of tumors. *Advances in Healthcare Materials* 9(14), 2000394.
75. Li, J., Liu, C., Hu, Y., Ji, C., Li, S., Yin, M. (2020) pH-responsive perylenediimide nanoparticles for cancer trimodality imaging and photothermal therapy. *Theranostics* 10(1), 166.
76. Xu, W., Jiao, L., Ye, H., Guo, Z., Wu, Y., Yan, H., Gu, W., Du, D., Lin, Y., Zhu, C. (2020) pH-responsive allochroic nanoparticles for the multicolor detection of breast cancer biomarkers. *Biosensors and Bioelectronics* 148, 111780.
77. Pei, X., Wu, X., Xiong, J., Wang, G., Tao, G., Ma, Y., Li, N. (2020) Competitive aptasensor for the ultrasensitive multiplexed detection of cancer biomarkers by fluorescent nanoparticle counting. *Analyst* 145(10), 3612–3619.

78. Qureshi, A., Tufani, A., Corapcioglu, G., Niazi, J.H. (2020) CdSe/CdS/ZnS nanocrystals decorated with Fe$_3$O$_4$ nanoparticles for point-of-care optomagnetic detection of cancer biomarker in serum. *Sensors Actuators, B Chemistry* 321, 128431.
79. Aydindogan, E., Ceylan, A.E., Timur, S. (2020) Paper-based colorimetric spot test utilizing smartphone sensing for detection of biomarkers. *Talanta* 2020 Feb 1;208:120446.
80. Kou, L., Sun, R., Jiang, X., Lin, X., Huang, H., Bao, S., Zhang, Y., Li, C., Chen, R., Yao, Q. (2020) Tumor microenvironment-responsive, multistaged liposome induces apoptosis and ferroptosis by amplifying oxidative stress for enhanced cancer therapy. *ACS Applied Materials & Interfaces* 12(27), 30031–30043.
81. Sohail, R., Abbas, S.R. (2020) Evaluation of amygdalin-loaded alginate-chitosan nanoparticles as biocompatible drug delivery carriers for anticancerous efficacy. *International Journal of Biological Macromolecules* 153, 36–45.
82. Chung, K., Ullah, I., Kim, N., Lim, J., Shin, J., Lee, S.C., Jeon, S., Kim, S.H., Kumar, P., Lee, S.K. (2020) Intranasal delivery of cancer-targeting doxorubicin-loaded PLGA nanoparticles arrests glioblastoma growth. *Journal of Drug Targeting* 28(6), 617–626.
83. Jogdand, A., Alvi, S.B., Rajalakshmi, P.S., Rengan, A.K. (2020) NIR-dye based mucoadhesive nanosystem for photothermal therapy in breast cancer cells. *Journal of Photochemistry & Photobiology, B: Biology* 208: 111901.
84. Mariadoss, A.V.A., Saravanakumar, K., Sathiyaseelan, A., Venkatachalam, K., Wang, M.H. (2020) Folic acid functionalized starch encapsulated green synthesized copper oxide nanoparticles for targeted drug delivery in breast cancer therapy. *International Journal of Biological Macromolecules* 164, 2073–2084.
85. Cherkasov, V.R., Mochalova, E.N., Babenyshev, A. V., Rozenberg, J.M., Sokolov, I.L., Nikitin, M.P. (2020) Antibody-directed metal-organic framework nanoparticles for targeted drug delivery. *Acta Biomaterialia* 103, 223–236.
86. Kang, J.K., Kim, J.C., Shin, Y., Han, S.M., Won, W.R., Her, J., Park, J.Y., Oh, K.T. (2020) Principles and applications of nanomaterial-based hyperthermia in cancer therapy, *Archives of Pharmacal Research,* 43, 46–57.
87. Van De Broek, B., Devoogdt, N., Dhollander, A., Gijs, H.L., Jans, K., Lagae, L., Muyldermans, S., Maes, G., Borghs, G. (2011) Specific cell targeting with nanobody conjugated branched gold nanoparticles for photothermal therapy. *ACS Nano* 5(6), 4319–4328.
88. Fazal, S., Jayasree, A., Sasidharan, S., Koyakutty, M., Nair, S. V., Menon, D. (2014) Green synthesis of anisotropic gold nanoparticles for photothermal therapy of cancer. *ACS Applied Materials & Interfaces* 6(11), 8080–8089.
89. Dadfar, S.M., Camozzi, D., Darguzyte, M., Roemhild, K., Varvarà, P., Metselaar, J., Banala, S., Straub, M., Güvener, N., Engelmann, U., Slabu, I., Buhl, M., Van Leusen, J., Kögerler, P., Hermanns-Sachweh, B., Schulz, V., Kiessling, F., Lammers, T. (2020) Size-isolation of superparamagnetic iron oxide nanoparticles improves MRI, MPI and hyperthermia performance. *Journal of Nanobiotechnology* 18(1), 1–13.
90. Marzella, L.L. (2018) NDA/BLA multi-disciplinary review and evaluation of SPY AGENT green, indocyanine green (NDA 211580 505(b)(2)). *Dmip/Ode Iv.* 505: 1–92.
91. Jiang, X., Du, B., Huang, Y., Yu, M., Zheng, J. (2020) Cancer photothermal therapy with ICG-conjugated gold nanoclusters. *Bioconjugate Chemistry* 31(5), 1522–1528.
92. Wang, X., Li, Y., Li, J., Gu, Y., Sang, L., Wang, D. (2020) NIR stimulus-responsive astragaloside IV–indocyanin green liposomes for chemo-photothermal therapy. *Journal of Drug Delivery Science and Technology* 64, 102257.

11 Nanobiotechnology in Agriculture

Pracheta Salunkhe, Chandni Krishnan,
Murthy Chavali, and Darshini Trivedi

11.1 INTRODUCTION

In the 21st century, nanobiotechnology has become prominent. The Greek prefix "nano," which means "dwarf," indicates a size of 10^{-9} and a combination of biomolecules at the nanoscale (Tomar et al., 2020).

The multidisciplinary field called "nanobiotechnology" integrates nanotechnology and biotechnology, and it is in its early stage of development. An industrial, medical, and agricultural green revolution is made possible by nanobiotechnology. It is an interdisciplinary field including biological sciences, surface science, organic chemistry, proteomics, pharmaceutics, molecular science, and semiconductor physics (Fakruddin et al., 2012; Baishakhi, 2022; Eliagh et al., 2021; Tripathi and Prakash, 2022).

Nanobiotechnology is applied in a variety of fields and has numerous applications. Agriculture is one of the most fascinating fields. A crucial part of the endeavor to meet the increased food demand of a fast-expanding global population is sustainable agriculture. One possible technique for enabling sustainable agriculture is nanobiotechnology. However, some nanoparticles (NPs) with special physiochemical properties naturally promote plant growth and stress tolerance rather than serving as nanocarriers (Zhao et al., 2020).

Modern smart, effective agricultural methods are driven by the use of nanobiotechnology concepts in agriculture. Agricultural nanobiotechnology mostly benefits from a plant nanobiotechnology technique that has demonstrated tremendous promise for targeted delivery and controlled release of agrochemicals, in addition to increasing the efficacy of agrochemicals. It may facilitate transgenic plant events, particularly in non-model crops. It can be utilized to increase stress tolerance, especially through early stress detection and sensing. Crop output may be increased via seed nano-priming (Chugh et al., 2021).

Compared to its usage in drug delivery and pharmaceuticals, nanotechnology has only recently been applied to the agriculture and food industries (Garcia et al., 2010). For "sustainable intensification," nanotechnology has the ability to boost global food production, improve food quality, detect plant and animal diseases, protect plants, monitor plant growth, and reduce waste (Gruère et al., 2011; Frewer et al., 2011; Pérez-de-Luque and Hermosin, 2013; Prasad et al., 2014; Biswal et al., 2012; Ditta,

DOI: 10.1201/9781003362258-11

2012; Sonkaria et al., 2012). One of the most critical areas of nanotechnology application is food and agricultural production (Coles and Frewer, 2013; Hong et al., 2013; Raliya et al., 2013; Kuzma and Verhage, 2014; Cortes-Lobos, 2011; Chen et al., 2014).

High reactivity, improved bioavailability and bioactivity, adherence effects, and surface impacts of nanoparticles are some of the characteristics and potential uses of nanotechnology in the agriculture sector. Nanotechnology is an exciting technology that may provide answers to many of the problems that people and the agriculture sector have been experiencing (Alvarado et al., 2019).

11.2   ROLE IN AGRICULTURE AND ITS APPLICATION

Agriculture has seen an increase in creative advancements in food security and productivity over the past few decades. The traditional agricultural industry uses agrochemicals such as fertilizers, pesticides, fungicides, insecticides, and herbicides to promote the growth and protection of crops. However, their excessive usage pollutes the ground and surface water, which has an impact on the ecology. By contrast, nanobiotechnology can alleviate agricultural problems and reduce the abuse of pesticides (Wu and Li, 2021; Dukare et al., 2016).

The application of insecticides is used to increase the shelf life of food products and the use of fertilizers increases crop cultivation. Nanobiotechnology is known for inventing and creating novel materials for use in a variety of fields. In light of these factors, nanotechnology offers significant benefits for the creation of new technologies tailored to agricultural needs. Through technology, nanoscience has demonstrated considerable promise for improved food quality, safety, and traceability, as well as for enhancing nutrient delivery, packaging performance, and agricultural and food processing (Resham et al., 2015).

Applications in the agriculture sector include: nano-fertilizers, nano-sensors, barcoding, nano-pesticides, and clay nanotubes (Figure 11.1).

a) Barcoding: This is used as identification, like tags. Progress in the field of nanoscale biotechnology has allowed for an increase in plant resilience to numerous environmental conditions including drought, salinity, diseases, and others. Through advancements in nanotechnology-based gene sequencing, it is anticipated that plant gene trait resources will soon be more effectively identified and used, enabling quick and affordable capabilities.

b) Nano-sensors: These are the sensors, utilizing advanced biosensor technology, which can provide efficient and precise means for the rapid identification of odors. Although the human nose can detect odors, there are still some microorganisms that the nose cannot detect. The main types are rapid detection biosensors, enzymatic biosensors, and electronic noses.

c) Nano-pesticides: The term "nano-pesticides" refers to preparations that intentionally introduce components in the nm size range. These nano-pesticides have already been introduced to the market and have unique properties connected to their small size range. Since they have been included in so many different goods, nano-pesticides cannot be considered a single category.

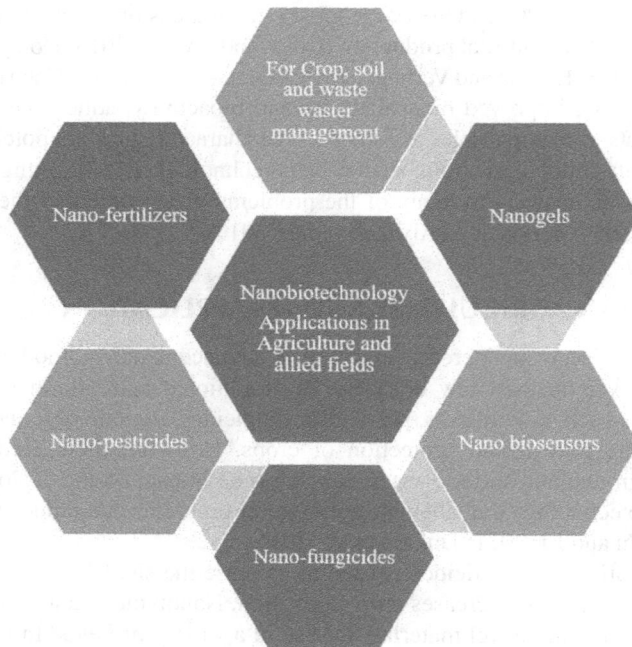

FIGURE 11.1   Schematic diagram depicting applications of nanobiotechnology in agriculture and allied fields.

d) Photocatalysis: Photocatalysis is one of the techniques that uses nanoparticles. It uses the terms "photo" and "catalysis," which together imply a "reaction generated by a catalyst." A huge increase in chemical reactivity and other physicochemical qualities connected to some specific situations, such as photocatalysis, photoluminescence, etc., occurs from the increase in surface atoms as the particle size decreases. Therefore, many harmful substances, including insecticides, which take a long time to disintegrate under normal circumstances, can be decomposed using this procedure. The main types are bioremediation of resistant pesticides, disinfectants, and wastewater treatment.

e) Clay nanotubes: Clay nanotubes (Halloysite) have been developed as pesticide carriers because of their low cost, longer release, and enhanced contact with plants. They can reduce the use of pesticides by 70–80%, which will reduce their cost and negative effects on water streams (Alvarado et al., 2019; Ditta, 2012; Malik et al., 2021; Kumari and Yadav, 2014).

f) Nanocomposites: Polymers that have been reinforced with small quantities (up to 5% by weight) of nanoparticles characterized by a high aspect ratio are known as nanocomposites. These materials can enhance the properties and performance of a polymer. Nanocomposites have a wide range of uses in many industries, including food packaging and agriculture (Kumari and Yadav, 2014).

11.3   ROLE IN INDUSTRIES AND ITS APPLICATION

Nanotechnology has been credited by the European Commission as one of the "Key Enabling Technologies" which has potential and can contribute significantly toward global sustainable competitiveness and numerous industrial sectors. With a focus on Sustainable Development Goals, a subset of nanotechnology, nanobiotechnology could play an important role in tackling challenges such as sustainability and food security by concentrating on the agriculture sector. The agriculture sector is in constant need of innovations to meet global food security needs which have been affected due to the climate change crisis. Major opposition to the use of nanobiotechnology in the agricultural sector is due to the uncertainty of its potential effects. Industries allied to the agriculture sector such as food processing and distribution, and food packaging could benefit from nanobiotechnology. Conventionally, the use of nanobiotechnology in agriculture has been focused on the production of improved plant varieties, disease management, crop protection, and in turn increased productivity. Even though the applications and advantages of nanobiotechnology are numerous, its actual contribution is observed to be small due to patent ownerships, low investment in economic interest returns, regulatory issues involved, and general public opinion/trust.

Some agro-products obtained by utilizing nanobiotechnology which can be commercialized include neem oil (*Azadirchta indica*) nano-emulsion as a larvicidal agent, macronutrient fertilizers coated with zinc oxide nanoparticles, soil enhancer products based on nano-clay components which are useful for water retention and release, filters coated with TiO_2 nanoparticles for photocatalytic degradation of agrochemicals in contaminated water, pesticide detection with a liposome-based nanobiosensor, and mesoporous silica nanoparticles which transport DNA to transform plant cells (plant genetic modification) (Parisi et al., 2015).

To meet the demands of sustainable agriculture and to control weeds and various pests, which include insects and fungi, researchers are working to find alternative solutions by using nanobiotechnology. These include nano-pesticides, nano-fertilizers, and nano-fungicides. The most common are nanoparticles used in agriculture including Ag, Fe, Cu, Si, Al, and Zn, ZnO, TiO_2, CeO_2, Al_2O_3, and most recently carbon nanotubes. Numerous nanomaterials are observed to possess antimicrobial properties which make them a suitable option as an antimicrobial agent in the food packaging industry. Nanoparticles are also used in food industries for developing foods which can have good nutritive value and also are of high quality. Nanobiotechnology is used to develop agrochemicals which focus on developing new delivery mechanisms to enhance crop productivity and reduce dependence on chemical pesticide application. A new role of nanobiotechnology in agriculture is the development of nano-sensors which aid in identifying potential diseases and residues of agrochemicals on crops (Figure 11.2). Nano-devices also have been created for genetic engineering of plants, plant disease diagnostics, and post-harvest management. In agriculture, and allied industries related to animal husbandry, nanobiotechnology is also used in poultry production, animal breeding, and improving animal health, and also in other sectors such as fisheries and aquaculture. As an agro- product nanobiotechnology is used to develop biomass for fuel production which can help to reduce the dependence on

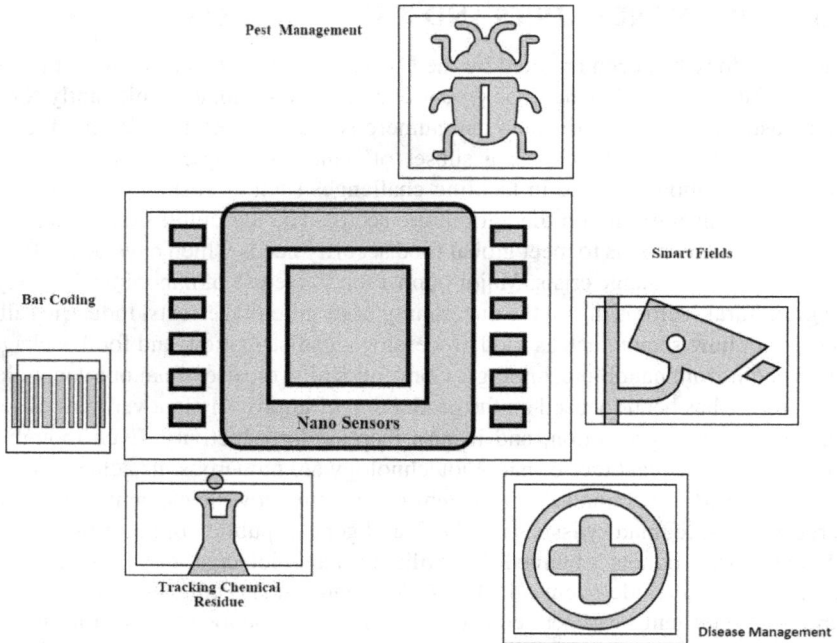

FIGURE 11.2 Schematic representations of the usage of nanosensors in the agriculture sector.

conventional fossil fuels. Nanobiotechnology is also used in wastewater treatment which can help plants in the absorption of nutrients. Nanobiotechnology also has wide application in the field of agriculture medicine (Ghidan and Antary, 2020).

Studies have been carried out to use nanoparticles as an innovative method for delivering bio-actives which could potentially control biotic and abiotic stress in plants. Also, the polymer industry can incorporate the use of nanobiotechnology in producing bionanocomposites which possess the desired properties of biopolymers such as biodegradability, bio-compatibility, and mechanical advantageous properties at the nanometric scale. Agro-products such as microbial nano-priming increase seed germination, vigor, rate of seedling emergence, and seedling growth and development. Nualgi Nano Biotech is a company situated in Bengaluru, India, which produces agro-products for increasing agricultural products.

In the agriculture sector, nanobiotechnology is an emerging field for the development of nano-pesticides and nano-fertilizers and it is evident that green synthesis of nanomaterials is preferred over chemical synthesis because it is less ecotoxic. Plant- or microbial-origin nanoparticle agro-products have shown potential to improve plant disease resistance and sustainable development of crops (Thakur, 2018).

Some of the novel agro-products and technologies used in incorporating nanobiotechnology in agriculture include nanoscale carriers, microfabricated xylem vessels, nano-lignocellulosic materials, clay nanotubes, photocatalysis, bioremediation of resistant pesticides, disinfectants, nano-barcode technology, smart dust, ZigBee, quantum dots, and biosensors which include the electronic nose, enzymatic, and rapid detection biosensors. Nanocarriers use the mechanism of entrapment and/or surface ionic and/or weak bond attachment, polymer and/or dendrimers, which are employed for efficient delivery of fertilizers, plant growth regulators, pesticides, and herbicides. Nano-lignocellulosic materials obtained from crops and trees include nano-whiskers innovated by Michigan Biotechnology Incorporate International which is focusing on using them as bio-composites which can substitute for fiber glass and plastic. Clay nanotubes (Halloysite) are also low-cost carriers of pesticides and, due to their efficiency, the amount of pesticides used is reduced by 70–80%, which in turn drastically reduces pesticide residue release into water streams. In photocatalysis, nanomaterials such as TiO_2, ZnO, SnO_2, and ZnS are used in the decomposition of toxic compounds used in agriculture. In the bioremediation of resistant pesticides, slow degradation and conversion to non-toxic compounds are carried out using nanoparticles. A disinfectant mechanism used is negative electron lone pairs which are generated due to the excitation of nanoparticles which when coming into contact with bacteria act as a disinfectant. Nano-barcodes are used as ID tags which can help in identifying those that are resistant to environmental stresses. Fluorescent labeled quantum dots with bio-recognition molecules have been used to detect various bacteria species affecting plants. Biosensors detect the volatiles produced by microbes and are utilized as rapid microbial detection tests. Smart dust used in vineyards and orchards monitors numerous parameters such as temperature, humidity, and in some cases insect and disease infestations. ZigBee is a wireless sensor technology using nanoscale sensitivity and has given rise to the concept of "Smart Fields" and "SoilNet" (Ditta, 2012).

The concept of "Sustainable Infestation" as per the Royal Society, London, aims to increase the agriculture yield in the same agricultural area without having an adverse environmental impact. An important factor to implement sustainable intensification in agriculture is to include nanotools in agriculture. Hydrogels, nano-clays, and nano-zeolites which enhance the water-holding capacity of soil by utilizing a mechanism where the slow release of water reduces hydric shortage period, especially during crop seasons, have proved to be favorable not only in agriculture but also in the reforestation of degraded areas (Fraceto et al., 2016).

Some commercially produced nano-fertilizers sold around the world include Nano-Gro manufactured by Agro Nanotechnology Corp., United States; Nano Green by Nano Green Sciences Inc., India; Nano-Ag Answer by Urth Agriculture, United States; Biozar Nano-Fertilizers by Fanavar Nano-Pazhoohesh Markazi Company, Iran; Nano Max NPK Fertilizers by JU Agri Sciences Pvt. Ltd, India; Master Nano Chitosan Organic Fertilizers by Pannaraj Intertrade, Thailand; and TAG NANO fertilizers by Tropical Agrosystem India (p) Ltd, India (Prasad et al., 2017).

Some examples of agri-food industries which have innovated consumer products utilizing nanobiotechnology available in the market are included in Table 11.1.

TABLE 11.1
Nanoproducts

Product name	Type	Manufacturer	Country
Canola Active oil	Cooking oil	Shemen Industries	Israel
Nanotea	Tea	Qinhuangdao Taiji Ring Nano-Products Co. Ltd.	People's Republic of China
Nanoceuticals Slim Shake Chocolate	Chocolate diet shake	RBC Life Sciences Inc.	US

Source: Sekhon (2014).

11.3.1 APPLICATIONS

In food technology, nanobiotechnology is used in food management where nanomaterials are used in nutraceuticals, as a form of nutrient delivery, for nutritional value enhancement due to mineral and vitamin fortification, nanoencapsulation of flavors, gelation, and viscosifying agents, to increase shelf life. Since nanobiotechnology in food technology is fairly new, studies related to bioavailability, absorption, distribution, metabolism, and excretion of nanoproducts need to be undertaken. Especially, bioaccumulation of nano-emulsions within the human body and their long-term effects should be studied. Regulatory bodies around the world which deal with safety issues related to nano-systems in food include the European Food and Safety Authority (EFSA), Environmental Protection Agency (EPA), Food and Drug Administration (FDA), National Institute for Occupational Safety and Health (NIOSH), Occupational Safety and Health (OSHA), US Department of Agriculture (USDA), and Consumer Product Safety Commission (CPSC) (Pradhan et al., 2015).

Nanobiotechnology's role in the food industry includes the customization of food in the form of alterations to the color and taste of food as per the dietary requirements, and major food manufacturing steps such as food processing, preservation, and packaging (Figure 11.3). Scientists are developing ways of incorporating nanobiotechnology for environmental remediation in which nanomaterials can decontaminate water for agricultural use. Nanomaterials have the potential to convert heavy metals or radioactive materials into less toxic compounds. Research is being carried out to understand the capacity of nanobiotechnology and in understanding the phenology and ecotoxicology and predicting its applications in respective fields (Gondal and Tayyiba, 2022).

Some of the applications of nanobiotechnology in agriculture can be broadly classified as smart monitoring and information management, smart disease prevention and control, smart pest control, the development of value-added products from animal by-products and wastes, and objective feeding and nutrition. The future scope for these applications include to explore safety dosages/limits, study physiochemical characteristics of agriculture and plant biota, and study their toxicity and the potential risks and bioremediation of nanomaterials itself (Vijayakumar et al., 2022).

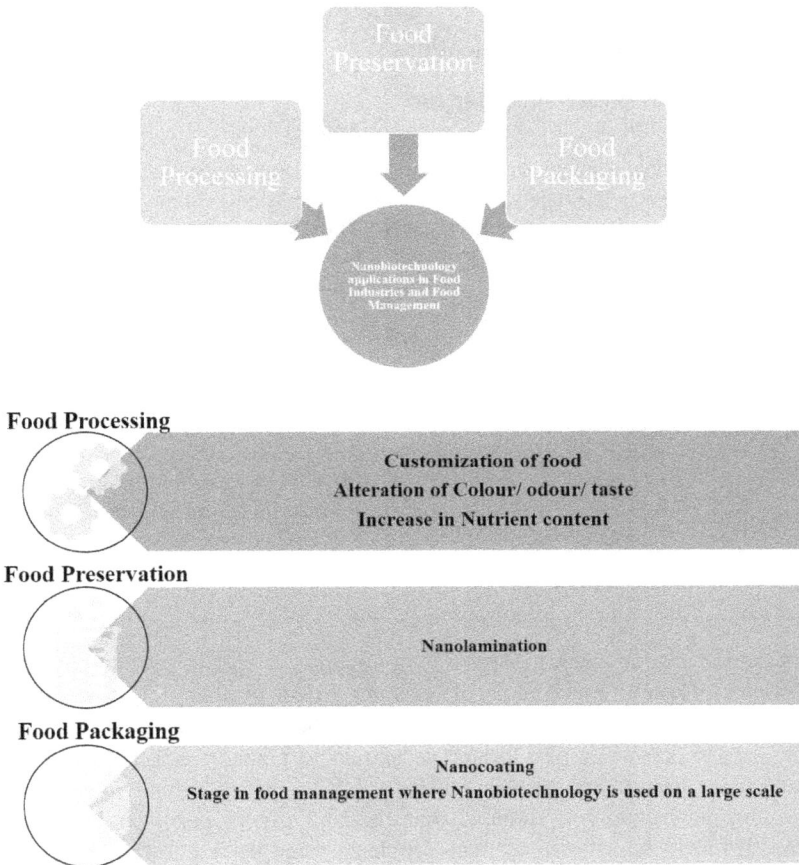

FIGURE 11.3 Schematic representation of applications of nanobiotechnology in food industries.

The use of agricultural drones combined with the usage of nanobiosensors is an emerging technology for calculation of the vegetable index, detection of infected plants, and assessing the number of chemicals required in the case of eradicating infestations. Nanobiotechnology is also used in toxin and adulterant detection roles in combination with other sophisticated analytical techniques (Halake and Muda, 2022).

There have been reports on the utilization of nanobiotechnology in common processes involved in the transformation of biomass to biofuels such as addition, pyrolysis, hydrogenation, transesterification, and gasification. Nanoparticles of calcium oxide and magnesium oxide have been successfully used in the transesterification of oil to biodiesel as biocatalyst carriers. *Camellia sinensis*, commonly called spent tea (solid waste), conversion to bio-ethanol by undergoing transesterification using bio-nanoparticles obtained from *Aspergillus niger* (Table 11.2) in a minimum of three steps can also be a source to produce alternative energy as throughout the world millions of

TABLE 11.2
Bionanoparticles and Their Applications

Nanoparticle	Biological source used for synthesis	Application
Au	*Trichoderma viride* (fungus)	Antimicrobial and enzyme immobilization
Ag	*Aspergillus niger* (fungus)	Wound healing (in rats)
Ag	*Lecanicillium lecanii* (fungus)	Antimicrobial textile cotton fabrics
Ag	*Trichoderma viride* (fungus)	In combination with sodium alginate increased the shelf life of carrots and peas
Au and Ag	*Fusarium oxysporum* (fungus)	Bioconjugate nano-PCR assay which is highly specific and sensitive
CuO	Gum carava, *Malva* sp., and *Phyllanthus* sp. (plant)	Antimicrobial against Gram-positive and Gram-negative bacteria
ZnO	*Camellia sinesis* (plant)	Antimicrobial
Ag	*Croton sparsiflorus* (plant)	Effective against Gram-positive and Gram-negative bacteria
TiO$_2$	*Oryza sativa* (plant)	Potential photocatalyst

Source: Thangadurai et al. (2020).

tons of tea are consumed. An interesting application of nanobiotechnology includes hydroponics, whereby metal nanoparticles are grown in plants. Particle farming is also one of the novel applications of nanobiotechnology in which researchers are working and successfully developing techniques to gather precious metals such as gold and silver from plants such as alfalfa.

The current scenario in India is that Indian Farmers Fertilizers Co-operative Limited (IFFCO), in order to mitigate problems faced by Indian farmers, has taken steps to promote novel agri-inputs. It has collaborated with Nano Biotechnology Research Centre (NBRC), Kalol, and Gujrat has successfully manufactured indigenous liquid nano-fertilizers such as nano urea, nano copper, nano zinc, and nano nitrogen. There have been multi-location trials conducted on multiple crops under varying crop seasons on these products and these nano-fertilizers have been food in sync with the regulatory requirements of OECD testing guidelines and Guidelines for Testing NAIPs and Food Products. Reports regarding the harvested crops that used nano-fertilizers are that they are fit for consumption, with no adverse effects and minimal environmental footprint (Nano for Agri, 2021).

Nanogels which immobilize pheromones (naturally occurring volatiles) have been reported as an efficient and eco-friendly biological pest control used in open orchards of guava (Sekhon, 2014). Nano-filtration is an important application where selective passing of particles is carried out for use in food industries to detect metabolites and other pathogenic factors and is also used as a quality control measure. Nano-filtration also has applications in wastewater treatment and water desalination (Roholla, 2011).

Nanolamination is a novel application in the food industry. It involves the protection of food from moisture, lipids, and gases. Nanolamination is a type of nano-coating that acts as a barrier to prevent deterioration of taste and color, and nutritional value loss, and also controls the growth of potential pathogens (Alvarado et al., 2019).

Nanobiotechnology has played an important role in genetically modifying Golden Rice; carbon nanofibers which contain foreign DNA material having the desired trait were used. Synthetic biology is a new branch of science which includes techniques of genetic engineering, informatics, and nanotechnology, and has helped researchers and scientists to develop novel plant varieties.

11.4 CONCLUSION

Nanobiotechnology is regarded as an innovative branch of research that has the potential for the advancement of sustainable industries including agriculture. The creation of materials with a nanometric scale (1–100 nm) that have a variety of properties depending on their size and manufacturing form, such as hydrophobicity, permeability, pH, etc., is one of these aspects. The development of nanodevices and nanomaterials may lead to new applications in plant biotechnology. The characteristics of nanoparticles utilized in agriculture have enhanced the targeted activity, with safe transportation with a reduced impact, and responded to specific plant conditions. Therefore, in order to enable their appropriate employment in the evolution of the processes that this sector will need in the future, it is imperative to carry out research on the potential toxicity of nanomaterials in agriculture. The importance of developing technological tools for agriculture should also be made known to researchers and investors in order to reduce the consumption of natural resources and the damaging impact that excessive chemical use has on the environment.

REFERENCES

Alvarado, K., Bolaños, M., Camacho, C., Quesada C. E., Vega-Baudrit, J. (2019) Nanobiotechnology in agricultural sector: Overview and novel applications. *Journal of Biomaterials and Nanobiotechnology* 10: 120–141.

Baishakhi, D., Goswami, T. (2022) *Nanobiotechnology – A Green Solution*. Wiley-VCH.

Biswal, S.K., Nayak, A.K., Parida, U.K., Nayak, P.L. (2012) Applications of nanotechnology in agriculture and food sciences. *International Journal of Science Innovations and Discoveries* 2(1): 21–36.

Chen, H., Seiber, J.N., Hotze, M. (2014) ACS select on nanotechnology in food and agriculture: A perspective on implications and applications. *Journal of Agricultural and Food Chemistry* 62(6): 1209–1212.

Chugh, G., Siddique, K., Solaiman, Z. (2021) Nanobiotechnology for agriculture: Smart technology for combating nutrient deficiencies with nanotoxicity challenges. *Sustainability* 13.

Coles, D., Frewer, L.J. (2013) Nanotechnology applied to European food production: a review of ethical and regulatory issues. *Trends in Food Science & Technology* 34(1): 32–43.

Cortes-Lobos, R. (2011) *Can Agri-Food Nanotechnology Contribute to Achieve the Millennium Development Goals in Developing Countries?* Los Polvorines, Argentina: Universidad Nacional de General Sarmiento. Available from: www.ungs.edu.ar/globelics/wp-cont ent/ uploads/2011/12/ID-15-Cortes-Learning-and-innovation-lessons-fromsectorial-studies.pdf. Accessed April 18, 2014.

Ditta, A. (2012) How helpful is nanotechnology in agriculture? *Advances in Natural Sciences: Nanoscience and Nanotechnology* 3(3): 033002.

Dukare, A., Bibwe, B., Nehru, B., Kadam, D. (2016) *Nanotechnology in Post Harvest Horticulture Management: A Review.* Chaudhary Charan Singh Haryana Agricultural University, Hissar, Haryana.

Eliajh, A., Musibau, A., Alao, M., Oke, M., Aina, A. (2021) Fungi as veritable tool in current advances in nanobiotechnology. *Heliyon* 7: e08480.

Fakruddin, M., Hossain, Z., Afroz, H. (2012) Prospects and applications of nanobiotechnology: a medical perspective. *Journal of Nanobiotechnology* 10: 31.

Faceto, L.F., Grillo, R., de Medeiros, G.A., Scognamiglio, V., Rea, G., Bartolucci, C. (2016) Nanotechnology in agriculture: Which innovation potential does it have? *Frontiers in Environmental Science* 4(Mar).

Frewer, L.J., Norde, W., Fischer, A.R.H., Kampers, F.W.H. (Eds.) (2011) *Nanotechnology in the Agri-Food Sector: Implications for the Future.* Weinheim, Germany: Wiley-VCH.

Garcia, M., Forbe, T., Gonzalez, E. (2010) Potential applications of nanotechnology in the agro-food sector. *Food Science and Technology (Campinas)* 30(3): 573–581.

Ghidan, Y.A., Antary, M.T. (2020) Applications of nanotechnology in agriculture. *Applications of Nanobiotechnology.* IntechOpen.

Gruère, G., Narrod, C., Abbott, L. (2011) *Agriculture, Food, and Water Nanotechnologies for the Poor: Opportunities and Constraints. Policy Brief 19.* Washington, DC: International Food Policy Research Institute. Available from www.ifpri.org/sites/default/files/publications/bp019.pdf. Accessed May 6, 2014.

Halake, H.N., Muda, H.J. (2022) Role of nanobiotechnology towards agri-food system. *Journal of Nanotechnology* 2022.

Hong, J., Peralta-Videa, J.R., Gardea-Torresdey, J. (2013) Nanomaterials in agricultural production: benefits and possible threats? In: Shamim, N., Sharma, V.K. (Eds.) *Sustainable Nanotechnology and the Environment: Advances and Achievements*, pp. 73–90. Washington, DC: American Chemical Society.

Kumari, A., Yadav, S. K. (2014) Nanotechnology in agri-food sector. *Critical Reviews in Food Science and Nutrition* 54(8): 975–984.

Kuzma, J., VerHage, P. (2006) *Nanotechnology in Agriculture and Food Production: Anticipated Applications.* Washington, DC: The Project on Emerging Nanotechnologies. Available from: www. nanotechproject.org/process/assets/files/2706/94_pen4_agfood.pdf. Accessed May 6, 2014.

Malik, A., Shayesta, I., Malik, M., Dar, Z. M., Masood, A., Shafi, S., Rashid, B., Sidique, S. (2021) *Nano Pesticides Application in Agriculture and their Impact on Environment.* Intech Open.

Nano for Agri (2021) *5th International conference on Nanobiotechnology for agriculture: Technology readiness for overcoming regulatory barriers to implement nanotechnology enabled agriculture for sustainable future.* Nano for Agri.

Parisi, C., Vigani, M., Rodríguez-Cerezo, E. (2015) Agricultural nanotechnologies: What are the current possibilities? *Nano Today* 10(2): 124–127.

Pérez-de-Luque, A., Hermosín, M.C. (2013) Nanotechnology and its use in agriculture. In: Bagchi, D., Bagchi, M., Moriyama, H., Shahidi, F. (Eds.) *Bio-nanotechnology: A Revolution in Food, Biomedical and Health Sciences*, pp. 299–405. Wiley-Blackwell: West Sussex, UK.

Pradhan, N., Singh, S., Ojha, N., Shrivastava, A., Barla, A., Rai, V., Bose, S. (2015) Facets of nanotechnology as seen in the food processing, packaging, and preservation industry. *BioMed Research International* 2015.

Prasad, R., Bhattacharyya, A., Nguyen, Q.D. (2017) Nanotechnology in sustainable agriculture: Recent developments, challenges, and perspectives. *Frontiers in Microbiology* 8(Jun).
Prasad, R., Kumar, V., Prasad, K.S. (2014) Nanotechnology in sustainable agriculture: present concerns and future aspects. *African Journal of Biotechnology* 13(6): 705–713.
Raliya, R., Tarafdar, J.C., Gulecha, K., et al. (2013) Review article; scope of nanoscience and nanotechnology in agriculture. *Journal of Applied Biology and Biotechnology* 1(03): 041–044.
Resham, S., Khalid, M., Gul, A. (2015) *Nanobiotechnology in Agricultural Development.* Springer.
Roholla, M.S., Rezaei, M. (2011) Nanotechnology in agriculture and food production. *Journal of Applied Environmental and Biological Science* 1(10): 414–419.
Sekhon, S.B. (2014) Nanotechnology in agri-food production: An overview. *Nanotechnology, Science and Applications* 7(2): 31–53.
Sonkaria, S., Ahn, S.H., Khare, V. (2012) Nanotechnology and its impact on food and nutrition: a review. *Recent Patents on Food, Nutrition & Agriculture* 4(1): 8–18.
Thakur, S. (2018) Bio-nanotechnology and its role in agriculture and food industry. *Journal of Molecular and Genetic Medicine* 12(1).
Thangadurai, D., Sangeetha, J., Prasad, R. (2020) *Nanotechnology for Food, Agriculture, and Environment.*
Tomar, R.S., Jyoti, A., Kaushik, S. (Eds.) (2020) *Nanobiotechnology: Concepts and Applications in Health, Agriculture, and Environment* (1st ed.). Apple Academic Press.
Tripathi, A., Prakash, S. (2022) *Nanobiotechnology: Emerging Trends, Prospects, and Challenges*. Woodhead Publishing Series.
Vijayakumar, M.D., Surendhar, G.J., Natrayan, L., Patil, P.P., Ram, P.M.B., Paramasivam, P. (2022) Evolution and recent scenario of nanotechnology in agriculture and food industries. *Journal of Nanomaterials* 2022: 1–17.
Wu, H., Li, Z. (2021) Recent advances in nano-enabled agriculture for improving plant performance. *The Crop Journal* 10.
Zhao, L., Lu, L., Wang, A., Zhang, H., Huang, M., Wu, H., Xing, B., Wang, Z., Ji, R. (2020) Nano-biotechnology in agriculture: Use of nanomaterials to promote plant growth and stress tolerance. *Journal of Agricultural and Food Chemistry* 68(7): 1935–1947.

Index

Note: Page numbers in **bold** refer to tables and those in *italic* refer to figures.

Taylor & Francis Group
an **informa** business

Taylor & Francis eBooks

www.taylorfrancis.com

A single destination for eBooks from Taylor & Francis
with increased functionality and an improved user
experience to meet the needs of our customers.

90,000+ eBooks of award-winning academic content in
Humanities, Social Science, Science, Technology, Engineering,
and Medical written by a global network of editors and authors.

TAYLOR & FRANCIS EBOOKS OFFERS:

A streamlined
experience for
our library
customers

A single point
of discovery
for all of our
eBook content

Improved
search and
discovery of
content at both
book and
chapter level

REQUEST A FREE TRIAL
support@taylorfrancis.com

Routledge
Taylor & Francis Group

CRC **CRC Press**
Taylor & Francis Group

For Product Safety Concerns and Information please contact our EU
representative GPSR@taylorandfrancis.com
Taylor & Francis Verlag GmbH, Kaufingerstraße 24, 80331 München, Germany

www.ingramcontent.com/pod-product-compliance
Lightning Source LLC
Chambersburg PA
CBHW070714220326
41598CB00024BA/3153